FAILED
TECHNOLOGY

TRUE STORIES OF TECHNOLOGICAL DISASTERS

FAILED TECHNOLOGY

TRUE STORIES OF TECHNOLOGICAL DISASTERS

Volume

2

**Fran Locher Freiman
& Neil Schlager**

*An imprint of Gale Research Inc.,
an International Thomson Publishing Company*

I(T)P

NEW YORK • LONDON • BONN • BOSTON • DETROIT • MADRID
MELBOURNE • MEXICO CITY • PARIS • SINGAPORE • TOKYO
TORONTO • WASHINGTON • ALBANY NY • BELMONT CA • CINCINNATI OH

FAILED
TECHNOLOGY
TRUE STORIES OF TECHNOLOGICAL DISASTERS

Fran Locher Freiman and Neil Schlager, *Editors*

Staff
Carol DeKane Nagel, *U·X·L Developmental Editor*
Thomas L. Romig, *U·X·L Publisher*

Mary Kelley, *Production Associate*
Evi Seoud, *Assistant Production Manager*
Mary Beth Trimper, *Production Director*

Mark Howell, *Page and Cover Designer*
Cynthia Baldwin, *Art Director*

H. Diane Cooper, *Permissions Associate (Pictures)*
Margaret A. Chamberlain, *Permissions Supervisor (Pictures)*

The Graphix Group, *Typesetter*

Library of Congress Cataloging-in-Publication Data
Freiman, Fran Locher
    Failed technology: true stories of technological disasters / Fran Locher Freiman, Neil
Schlager.
       2 v. (xviii, 392 p.): ill.; 24 cm.
    Includes bibliographical references (v. 1, p. 173-175, v. 2, p. 383-385) and indexes.
    ISBN 0-8103-9794-3 (set). -- ISBN 0-8103-9795-1 (v. 1). -- ISBN 0-8103-9796-X (v. 2).
    1. System failures (Engineering)--Case studies. 2. Disasters--Case studies. I. Freiman, Fran Locher II.
Schlager, Neil, 1966- III. Title.
TA169.5 .F74 1995
620.20--dc20

96-129100
CIP

∞™ This book is printed on acid-free paper that meets the minimum requirements of American National Stan-dard for Information Sciences—Permanence Paper for Printed Library Materials, ANSI Z39.48-1984.

Printed in the United States of America
10 9 8 7 6 5 4 3 2

Contents

Preface

People today can get very enthusiastic about using technology, especially because it mostly works very well and so greatly enhances the quality of our lives. Technological tools are around us everywhere—they wake us up, entertain us, keep our food fresh, help us prepare and cook our meals and wash the dishes afterwards. They take us where we want to go and keep us comfortable while we are on our way. We look forward to the new and improved technologies, and we hope they come in our favorite color and in the latest style.

Our expectations about the technological devices in our lives are so high because we rely on them so completely. We hardly question how all these tools came to exist, and this reliance itself testifies to how well technology works most of the time. But we have all experienced occasions of failed technology, such as when the electrical power goes out or our cable television company has technical problems. We hate being inconvenienced.

But what about when the failure is a major one? Henry Petroski writes in his new book, *Design Paradigms*, "Some of history's most embarrassing moments have come in the dramatic failures of some of the largest machines, structures, and systems ever attempted." It surely was embarrassing to the owners of the steamship *Titanic* when the liner sank, despite the owners' boast that their ship was unsinkable. And NASA was certainly embarrassed when the space shuttle *Challenger* exploded and when the shuttle inquiry revealed the embarrassing facts about the O-ring deficiencies.

However, failed technologies produce far more than embarrassment. They endanger human lives or end them. They disrupt our economic livelihoods. They shock us and set us back in our confidence. They make us question how we came to rely so much on manufactured products. Failures frighten us.

But aside from what we lose in disasters, we also learn from them. All technological development represents the thoughtful application of design expressed in materials. When successful designs are modified, the new design seems improved. However, Petroski reminds us: "Any design change . . . can introduce new failure modes or bring into play latent failure modes. . . . While a structure designed 'the old way' may be perfectly safe, an 'improved' or enlarged design could hold very unpleasant surprises."

Format and Inclusion Criteria

Failed Technology: True Stories of Technological Disasters focuses on 44 unpleasant surprises of the twentieth century. The failed technologies span several fields:

- Ships and Submarines
- Airships, Aircraft, and Spacecraft
- Automobiles
- Dams and Bridges
- Buildings and Other Structures
- Nuclear Plants
- Chemical and Environmental Disasters
- Medical Disasters.

The catastrophes highlighted in *Failed Technology* resulted from poor design, planning, testing, or construction—not from purely natural disasters such as earthquakes and hurricanes nor deliberate human actions such as terrorist bombings. Other disasters are included because of the media coverage they received and because of their impact on public opinion. Many of the failures are well known. But even when they are not dramatic or close to home, they reveal how technology has gone wrong.

The entries in *Failed Technology* have a common format. Each contains major sections in the following pattern:

- Background
- Details of the Disaster
- Impact
- Where to Learn More

Additional headlines highlight the specific incidents that led up to the catastrophe and occurred in its aftermath. The volumes also feature more than 70 photos and illustrations, a chronology of technological disasters, a bibliography, and an cumulative subject index.

Comments and Suggestions

We welcome your comments on this work as well as your suggestions for disasters to be featured in future editions of *Failed Technology: True Stories of Technological Disasters*. Please write: Editors, *Failed Technology*, U·X·L, 835 Penobscot Bldg., Detroit, Michigan 48226-4094; call toll-free: 1-800-877-4253; or fax: 313-961-6348.

Picture Credits

The photographs and illustrations appearing in *Failed Technology: True Stories of Technological Disasters* were received from the following sources:

The Bettmann Archive: pages 4, 6, 74, 88; **AP/Wide World Photos:** pages 9, 25, 29, 39, 54, 118, 129, 134, 136, 150, 155, 162, 186, 193, 198, 227, 237, 251, 261, 266, 276, 302, 315, 319, 321, 361; **UPI/Bettmann Newsphotos:** pages 13, 64, 284, 286, 342; **UPI/Bettmann:** pages 20, 61, 71, 79, 96, 103, 110, 206, 209, 214, 222, 225, 248, 255, 257, 334, 335, 345, 377; **Reuters/Bettmann:** pages 23, 91, 126, 300; **U.S. Navy:** pages 36, 44; **Archive Photos/Lambert:** page 69; **Reprinted by permission of W. W. Norton & Company, Inc.:** page 81; **Copyright 1979 Time Inc., reprinted by permission:** p. 105; **Sovfoto:** page 141; **NASA:** pages 147, 167, 169; **Kyodo News Service:** page 294; **Archive Photos/Orville Logan Snider:** page 312; **Black Star:** page 327; **Peter A. Simon/Phototake, NYC:** page 370.

Chronology of Technological Failures

September 30, 1911	Austin Dam fails
April 15, 1912	R.M.S. *Titanic* sinks
September 3, 1925	U.S.S. *Shenandoah* crashes
October 5, 1930	R-101 crashes
May 6, 1937	*Hindenburg* explodes
May 23, 1939	U.S.S. *Squalus* sinks
1939—	DDT insecticide contamination
November 7, 1940	Tacoma Narrows Bridge collapses
1940-79	Diethylstilbestrol (DES)
1942-80	Love Canal toxic waste site
January 10, 1954	BOAC Comet explodes
April 8, 1954	BOAC Comet explodes
1955—	Minamata Bay mercury poisoning
July 26, 1956	*Andrea Doria* sinks
September 29, 1959	Lockheed Electra crashes
1959-63	Chevrolet Corvairs roll over
1950s-60s	Thalidomide
March 17, 1960	Lockheed Electra crashes
December 16, 1960	United Airlines DC-8 and TWA Constellation collide
January 3, 1961	SL-1 reactor explodes
1961-71	Agent Orange contamination
April 10, 1963	U.S.S. *Thresher* sinks

October 9, 1963	Vaiont Dam landslide
January 27, 1967	Apollo 1 catches fire
April 24, 1967	*Soyuz 1* crashes
May 27, 1968	U.S.S. *Scorpion* lost
1960s—	Silicone-gel implants
April 13, 1970	*Apollo 13* oxygen tank explodes
1971-76	Ford Pintos explode (1971-76 model years)
May 14, 1973	Skylab's meteoroid shield fails
March 28, 1979	Three Mile Island reactor melts down
May 25, 1979	American Airlines DC-10 crashes
1970s	Dalkon Shield Intrauterine Device
November 21, 1980	MGM Grand Hotel fire
March 8, 1981	Tsuruga radioactive waste spills
July 17, 1981	Hyatt Regency Hotel walkways collapse
February 15, 1982	*Ocean Ranger* rig sinks
August 1982	Zilwaukee Bridge fails
December 3, 1984	Union Carbide toxic vapor leak
August 12, 1985	Japan Airlines Boeing 747 crashes
January 28, 1986	*Challenger* explodes
April 26, 1986	Chernobyl reactor explodes
April 5, 1987	Schoharie Creek Bridge collapses
February 24, 1989	United Airlines Boeing 747 explodes
March 24, 1989	Exxon *Valdez* runs aground
October 4, 1992	El Al Boeing 747-200 crashes

III

Automobiles

Chevrolet Corvairs roll over

1960–63 model years

Background

The Chevrolet Corvair, introduced in September 1959, was the first mass-produced rear-engine automobile in the United States. It was involved in a series of accidents, some of which were fatal, which were blamed on flawed suspension design. General Motors (GM), its manufacturer, faced a flurry of lawsuits. The Corvair's direct-air heating system was also charged with drawing noxious and potentially dangerous fumes into the passenger compartment. The controversy surrounding the Corvair, in which consumer advocate Ralph Nader gained national prominence, brought about the passage of the 1966 National Traffic and Motor Vehicle Safety Act, the first federal legislation ever to mandate standards for automotive design safety.

Alleged flaws in the Chevrolet Corvair's rear suspension causes accidents—some fatal—and brings the American automotive industry under federal regulation.

Rear-mounted engine stirs controversy

The Corvair was a compact car whose design was quite radical—by the standards of the conservative American automotive industry. Its rear-mounted engine was air-cooled like that of the Volkswagen. Unlike its imported rival, however, the Corvair featured an innovative engine and transaxle system. It had fully independent coil spring suspension, and its unit-body construction nearly eliminated the heavy iron frame of traditional American automobiles.

The Corvair's design generated controversy even before the car was introduced to the public. No rear-engine car had ever been mass-produced in America, and there were many skeptics. In fact, in advance advertisements for the Ford Falcon and the Chrysler Valiant—the models

In this 1959 Ford-produced film, a Corvair (right) skids out of control while trying to keep up with a Ford Falcon. Consumer activist Ralph Nader charged that the film proved wrong General Motors' claim that the Corvair had good handling ability.

Ford and Chrysler were introducing at this time—GM's competitors tried to raise doubts about the stability of a rear-engine compact car. Magazine reviewers were more positive in tone, but they still raised questions about the car's high-speed handling. Nonetheless, the Corvair initially enjoyed respectable sales, and in April 1960 *Motor Trend* magazine named it "Car of the Year"—a coveted industry honor.

Details of the Controversy

The first documented fatality in a Corvair involving loss of control occurred in 1960, when a California teenager named Don Lyford crossed the centerline of the road during an S-turn and was killed in a collision with an oncoming car. A Los Angeles attorney investigating the accident interviewed a police officer. The officer claimed to have seen "six of them [Corvairs] flip out of control" in recent months.

GM settles with woman who lost arm in rollover crash

Over the next several years, this attorney litigated scores of Corvair legal cases against General Motors. None of these cases reached a definite conclusion until 1964, when Rose Pierini, who had lost an arm in a rollover crash, was awarded $70,000 in an out-of-court settlement. This settlement was widely reported in the press as a victory for the plaintiff, and it opened a legal floodgate against GM.

Published in November 1965, Ralph Nader's *Unsafe at Any Speed* is popularly regarded as the book that killed the Corvair. Only the first chapter actually discusses the Corvair, but Nader intended the book to be an indictment of all the "sins" of the Detroit automobile establishment, and principally of General Motors. The first sentence of the preface reads: "For over half a century, the automobile has brought death, injury, and the most inestimable sorrow and deprivation to millions of people." Nader contended that General Motors executives marketed the Corvair even though they knew it was unsafe, because their desire for profit outweighed all other considerations. Nader stopped just short of accusing the automaker of negligent homicide.

Ralph Nader plays major role in getting national safety law passed

Under ordinary circumstances *Unsafe at Any Speed* might have remained an obscure publication. But it received widespread attention because Senator Abraham Ribicoff chaired the Senate subcommittee that was drafting a bill to establish federal standards for automotive design. Nader was working on Ribicoff's staff, and during congressional hearings he testified as an expert witness on various automotive topics, including the safety of the Corvair.

Nader's testimony and some well-timed publicity set the stage for the National Traffic and Motor Vehicle Safety Act to be passed unanimously by both houses of Congress. President Lyndon B. Johnson signed it into law September 9, 1966. Nader's role in the passage of the bill established him as a kind of "patron saint of consumerism." He had to compromise somewhat from his original demand that GM initiate a total recall of all 1960–63 Corvairs—the new legislation did not provide for that. In fact, desspite nearly ten years of effort by Nader toward this end, a comprehensive recall of the Corvair never took place.

GM responds—grudgingly—to heater-vent complaints

General Motors discontinued production of the Corvair in 1969, but

Ralph Nader's Unsafe at Any Speed is popularly regarded as the book that killed the Corvair. Only the first chapter actually discusses the Corvair, but Nader intended the book to be an indictment of all the "sins" of the Detroit automobile establishment, and principally of General Motors.

Chevrolet's litigation problems continued. A new round of allegations now concerned the design of the vehicle's direct-air heater, which returned public attention to the beleaguered Corvair. Over the decade GM received hundreds of letters complaining of acrid—and even dangerous—fumes emitted through the heater vents. Consumer advocate groups exerted continual pressure on the National Highway Traffic Safety Administration, which finally determined that the Corvair's direct-air heating system "creates an unreasonable risk of accidents and injury to persons. . . . Such engine fumes do in some cases contain carbon monoxide in sufficient concentrations to harm or endanger the occupants of the vehicle."

General Motors reacted grudgingly to this finding. In December 1971 GM mailed letters to all known Corvair owners and suggested that they have the heating system inspected at a dealership. The automaker claimed that the release of fumes depended on the condition of engine and heater duct seals, which GM considered a maintenance matter for which it had no financial responsibility. The owner would have to pay for complying with the recommended inspection. GM then suggested that winter drivers turn the heater off or roll the windows down if the fumes became too noxious—a suggestion that was not well-received. But General Motors, already embattled over the Corvair's suspension system, resolved to wait out this new tempest, because the cars were aging and would be removed from service in due time.

Details of the Corvair's Design

The Corvair's design engineers tried to slim down its powertrain (the system through which power, generated by the rear-mounted engine, is transmitted to the axle). Nevertheless, more than 60 percent of the vehicle's weight remained concentrated on the rear axle. The Corvair had fully independent suspension (each wheel could move without affecting the movement of any other wheel), and the rear wheels were driven by "swing axles" between the wheels and transaxle. These axles had universal joints only on the side connected to the transaxle.

The question is: Can the average driver cope with car's sideways thrust?

Most cars "understeer" during normal driving, and in a high-speed turn their tendency is to go straight—the driver must increase the rotation of the steering wheel to compensate. An oversteering car tends to increase the angle of a turn—the driver must adjust by reducing the rotation of the steering wheel.

One of the complaints against the Corvair was that, because the weight of the engine was in the back of the car, its front (that is, steering) wheels were lightly loaded. The vehicle would understeer during normal driving—as most cars do—but suddenly begin to oversteer during a high-speed turn and generate significant lateral (sideways) thrust. This phenomenon could get an unskilled and unprepared driver into trouble. The dispute was not over whether the phenomenon existed—even the Corvair's defenders did not deny that it did—but under what conditions the oversteering occurred, and whether the average driver could be expected to cope with it.

One strategy adopted by the Corvair's designers to offset the over-steering phenomenon was to stipulate that different air pressures be maintained in the front and rear tires. The specified cold pressure in the front tires was 15 psi (pounds per square inch), and for the more heavily loaded rear tires, 26 psi. Although these instructions were printed in the owner's manual, Corvair owners routinely ignored (or were unaware of) them, and anyone seeking advice on the subject at service stations often received incorrect and contradictory information. The failure to maintain the tire-pressure differential exacerbated the car's tendency to oversteer.

Nader claims that Corvair's suspension increases rollover tendency

The other allegation regarding the Corvair's suspension system related to the design of the rear drive axles. Nader and other critics claimed that during sharp cornering, the camber (vertical angle) of the inboard rear wheel could shift by as much as 11 degrees. Since there was no universal joint between the wheel and the axle, this purported angle change would be transmitted to the axle. Acting as a lever, the axle would then exert an upward force on the transaxle, causing excessive body sway and greatly increasing the likelihood that the car would roll over. Critics made much of the fact that a front anti-roll bar was originally included in the Corvair design but had been eliminated to reduce costs.

General Motors fervently denied the allegation about the camber of the inboard rear wheel, and the Corvair's critics never proved it. However, there was some evidence to support the claim. Shortly after the introduction of the Corvair, aftermarket accessories became available to improve the car's ride and handling. The best-selling of these, the EMPI Camber Compensator, was widely touted as a cure for the car's suspension ills. In the 1964 model year GM adopted suspension modifications very similar to the camber compensator, after which the incidence of accident-related lawsuits decreased dramatically.

Sales of "spin, flip, and burn" car drop drastically

In the 1965 model year the Corvair's rear suspension was completely redesigned. By that time, however, negative publicity about the car had done its damage. The Corvair's sales revived slightly in 1965, but in 1966, the year of the Ribicoff hearings, sales dropped catastrophically from 209,152 to 88,951. Virtually no model changes occurred between 1966 and 1969, the car's final year of production, and the Corvair was officially canceled on May 14, 1969. In that year, only 3,102 Corvairs were produced.

The complaints about the Corvair's heating system dogged the car from the start, but they were overshadowed by the suspension controversy. In most cars the passenger compartment heater relies on heat drawn from the engine coolant. The Corvair—with an air-cooled engine mounted in the rear—relied on ducts and an electric fan to route some of the air used to cool the engine into the passenger compartment. If noxious fumes such as oil, smoke, or carbon monoxide escaped from worn engine seals—a virtual certainty, given the state of the technology—they were pulled into the passenger compartment.

Impact

The outcome of the Corvair controversy was a new legal concept: the liability of an automobile manufacturer for negligence of design safety. Until then vehicle interiors, for example, had no seat belts or interior padding, headrests, or knee bolsters. In hindsight, protruding knobs and horn buttons often produced injuries that would have been easily preventable; the same was true of protruding fins and hood ornaments on the exterior. Brake systems were not fail-safe, steering columns sometimes broke loose and impaled drivers, and broken windshields were deadly.

New ethic arises: Product safety and manufacturer responsibility

The negative publicity surrounding the Corvair galvanized a new attitude about product safety and manufacturer responsibility. The National Traffic and Motor Vehicle Safety Act developed by Senator Ribicoff's subcommittee was largely written by Ralph Nader, who regarded auto manufacturers as criminals. Nader regarded the Safety Act adopted into law as not nearly strong enough (for example, he held that executives should be imprisoned for noncompliance). Nevertheless it brought into existence the National Traffic Safety Agency, later called the National Highway Traffic Safety Administration, which drafts the regulations adhered to by all auto manufacturers today.

Ironically, that agency issued a report in July 1972 that exonerated the Corvair of charges that it was exceptionally unstable in handling. The report concluded, "The handling and stability performance of the 1960–1963 Corvair does not result in an abnormal potential for loss of control or rollover, and it is at least as good as the performance of some contemporary vehicles both foreign and domestic."

Where to Learn More

"Car of the Year." *Motor Trend* (April 1960).

"Corvair Corsa Road Test." *Motor Trend* (January 1965).

Cray, Ed. *Chrome Colossus: General Motors and Its Times.* New York: McGraw-Hill, 1980.

Knepper, Michael. *The Corvair Affair.* Osceola, WI: Motorbooks International, 1982.

McCarry, Charles. *Citizen Nader.* N.p.: Saturday Review Press, 1972.

"Miscellaneous Ramblings." *Road & Track* (October 1966).

"My Six Years in the Spin, Flip, and Burn Car." *Car Collector* (July 1981): 17–19.

Nader, Ralph. *Unsafe at Any Speed.* New York: Grossman, 1965.

National Highway Traffic Safety Administration. "Evaluation of the 1960–1963 Corvair Handling and Stability." *DOT HS-820.* Washington, DC: Government Printing Office, July 1972.

O'Connell, Jeffrey, and Arthur Myers. *Safety Last: An Indictment of the Auto Industry.* New York: Random House, 1966.

"Safe at Any Speed." *Road & Track* (July 1966).

"SCI Analyzes Ed Cole's Corvair." *Sports Car Illustrated* (November 1959): 23.

Sobel, Robert. *Car Wars.* New York: Dutton, 1984.

Ford Pintos explode

1971–76 model years

At least 59 people are fatally burned when their cars burst into flames in rear-end collisions. Ford Motor Company is pressured to recall defective autos and submit to government regulation.

Background

Soon after its debut in 1971, the Ford Pinto became the hottest-selling subcompact on the market. Over its 1971–80 production life span, more than 3 million Pintos were sold. By the middle of the decade, however, reports proliferated that the Pinto tended to catch fire in rear-end collisions. In 1978, under pressure from the National Highway Traffic Safety Administration, Ford recalled 1.4 million 1971–76 Pintos for safety modifications. At least 59 people had burned to death in Pinto accidents by the mid-1980s, and more than 100 lawsuits forced the automaker to pay out many millions of dollars in damages, including $6.5 million plus interest awarded to Richard Grimshaw, who was horribly burned in May 1972.

Pinto succeeds as "import-fighter"

Disproving the widely held assumption that small cars do not make money, the Pinto sold 250,000 units in its first 12 months—which was good news for Ford Motor Company and Lee Iacocca. Then executive vice-president of Ford Cars and Trucks, Iacocca spearheaded the Pinto as an "import-fighter." Iacocca's goal was to market a sensible "econocar" priced under $2,000 and weighing less than 2,000 pounds.

Not long afterwards the Center for Auto Safety, a group run by disciples of consumer activist Ralph Nader, reported that it was receiving "significant numbers" of alarming complaints about the Pinto. It seemed that Pinto fuel tanks punctured easily in rear-end crashes, setting off a fireball that killed or seriously burned the car's occupants. The Center for Auto Safety demanded government action. In fall 1977 the National Highway

Wreckage of a 1973 Ford Pinto, which burst into flames when another car rear-ended it. One person died in the July 1978 crash, one month after Ford recalled 1.4 million Pintos built in 1971–76.

Traffic Safety Administration (NHTSA) began its "defects investigation" on the Pinto fuel-tank fires.

The public wants small cars

Ford had a reputation in the mid-1950s as a safety pioneer with such innovations as seat belts. In the late 1960s, however, the emphasis in the automobile market had shifted considerably. Losing market share to smaller imported cars was a real wake-up call to Lee Iacocca at Ford. Iacocca had convinced Henry Ford II to produce the sporty Mustang, and sales had been phenomenal: Over 1 million units sold during 1965–66, a whopping 78.2 percent of all the cars sold in North America in the small-sporty segment.

If people were ready to buy small cars, Iacocca was ready to deliver products to them. After convincing Ford management—"which . . . effectively meant [convincing] Henry Ford II," as Robert Lacey recounted in his book *Ford: The Men and the Machine*—Iacocca tore through the various production phases at breakneck speed. But producing a totally new car, which the Pinto was, posed engineering challenges regarding every product objective, not just safety. And it required time.

Mistakes made in Pinto design were serious

One of the issues for this early domestic entry into the small-car market was gas tank design. According to a 1969 memo, the automaker rejected an over-the-axle gas tank design because it would have slashed luggage space. The fuel tank chosen for the Pinto preserved luggage space, even though the over-the-axle design was inherently safer.

The typical early development phase—design, mock-up, and testing—for new automobile models at that time averaged 43 months. Iacocca developed the Pinto in 37 months. As Lacey noted, "The rush programme for the car's testing, production, and launch simply underestimated, or refused seriously to consider, the time, cost, and technical problems involved in such an ambitious undertaking." Tooling alone, which started before quality assurance was completed, cost Ford $200 million.

"The down side" of shaving normal development schedules, Lacey wrote, "was that in all the haste and improvisation, mistakes could creep into the final car—and in the case of the Ford Pinto, these mistakes were to prove serious."

Details of the Controversy

Soon after the Pinto was launched, Lily Gray and her passenger, Richard Grimshaw, were merging onto a Santa Ana, California, highway, when the engine in her new Ford Pinto stalled. Struck from behind, the Pinto was seriously damaged and the gas tank was ruptured. When a spark ignited the fuel vapors, the Pinto's occupants were engulfed in flames. "Lily Gray died horribly a few hours later, in a nearby hospital's emergency room," wrote Lacey, "her entire body having been charred and incinerated in the explosion. Her . . . passenger, Richard Grimshaw, a thirteen-year-old schoolboy, lingered on, scarred inhumanly, his features literally melted away."

Safety doesn't sell

Lee Iacocca, who would later write in his memoirs that he accepted full responsibility for the Pinto, declared, "Believe me, nobody sits down and thinks, 'I'm deliberately going to make this car unsafe.'" But the industry wasn't concerned with safety. As Lacey related, Henry Ford II said in 1969, "Look, . . . we could build the safest car in the world. We could build a tank that would creep over the highway and you could bang 'em into each other and nobody would ever get a scratch. But nobody would buy it either. . . . The American people want good cars, good looking cars, fast cars, cars with power and styling, and that's the kind of cars we build." Ford added, in reference to Ralph Nader and his Center for Auto Safety, "We spend a hell of a lot of time trying to make [cars] better and safer, and then some pipsqueak who doesn't know a thing about the industry, comes along and tries to tell us how to do what we've dedicated our lives and billions of dollars to doing."

But the fact was that Nader, who, Lacey conceded, "did overstate his case on occasions," conducted a successful campaign to give safety more importance. "As a consequence," Lacey wrote, "[automakers] were, by the beginning of the 1970s, being subjected to government regulation of the most detailed, and, in their view, most hindersome sort."

Mother Jones exposé claims Ford knew about flaws

A lengthy exposé by Mark Dowie in *Mother Jones,* a San Francisco-based investigative magazine, in fall 1977, enumerated the problems of the Ford Pintos. Dowie criticized the manufacturer for cutting the Pinto's development phase short, for using a fuel tank with structural flaws, and for positioning it in a particularly vulnerable location. Dowie also insisted that Ford knew about the flaws. According to previously concealed corporate documents, Dowie contended that Ford executives knew from their own testing that Pintos were prone to catch fire after rear-end collisions. Dowie denounced Ford for opting not to correct the problem because the repairs—at an estimated $10 to $15 per car—would cost more than Ford thought it would pay out in accident-related lawsuits.

Dowie highlighted three design details:

- The tank was mounted far to the rear, barely six inches ahead of the bumper.

- The fuel filler tube was prone to separate and create spillage.

- Four bolts protruded backward from the differential housing; during a rear-end collision, the bolts could easily puncture the tank, which was only three inches away.

Several other under-chassis components could pierce the fuel tank, Dowie charged. He cited other flaws: The Pinto's floor pan could separate and

The rear axle is the shaft on which the rear pair of wheels revolves. The differential housing is the case enclosing the differential gears, which connect two shafts or axles, permitting one shaft to revolve faster than the other.

allow a fuel tank fire to spread into the passenger compartment. The vehicle's front doors tended to jam shut in a crash and prevent escape. The Pinto had flimsy bumpers and generally fragile construction. And the Pinto was the first modern American Ford to lack rear subframe members. As Lacey explained, rear subframe members are "the solid steel skeleton which both carries the sheet metal of a conventional rear trunk and also protects the fuel tank in the event of a rear-end collision."

Rear-ended Pinto "would buckle like an accordion"

In a rear-end collisions at 30+ mph (miles per hour), claimed *Mother Jones*, a Pinto's rear end "would buckle like an accordion, right up to the back seat. The tube leading to the gas-tank cap would be ripped away from the tank itself, spilling the gas all over the road." The buckled tank would jam against the differential housing, against its protruding bolts. Add a random spark, and both cars "would be engulfed in flames." Safety expert Byron Bloch called it "irresponsible" to put "such a weak tank in such a ridiculous location in such a soft rear end."

Ford added more steel to Pinto rear end

"Ford soon realised its mistake," Lacey wrote. "Partly to protect the Pinto fuel tank, it started putting steel—and hence weight—back into the rear of the car." Lacey acknowledged that "some of these weight increases came from the addition of creature comforts. But some represented the difference between life and death. By the end of the Pinto's production life, Ford engineers had made it a reasonably safe and roadworthy vehicle, but they had achieved this by totally abandoning Iacocca's original 2,000-pound target."

Ford documents in fact revealed that early Pintos had been tested more than 40 times. Engineers reported that in a 21.5 mph test, the "fuel filler pipe pulled out of the fuel tank and the fluid discharged through the outlet. Additional leakage occurred through a puncture." Noted *Mother Jones*, "Every test at twenty-five miles per hour or more . . . resulted in a ruptured fuel tank." Ford's testing continued after the Pinto was in production. A report on fuel tank integrity of 1971–72 Pintos concluded that the rear-end structure was unsatisfactory for a 20-mph crash. Major revision would have to be undertaken in order for the Pinto to withstand a 30-mph collision.

Ford documents also contained proposals to improve the Pinto's rear-end safety. Possible fixes included repositioning the tank, moving the filler pipe, shielding the protrusions, installing a bladder in the tank, and—the most effective of the solutions—adopting a fuel tank that rides above the rear axle and differential. Ford actually held a patent on that type of tank and used it on its European-built Capri, which it imported into the United States. Soon after the Pinto's debut, the Capri passed the equivalent of a 45-mph rear-end crash test. British ads boasted about the Capri tank's safety.

Pinto meets federal standards of the time

Ford argued that the Pinto's fuel system presented no unreasonable risk. The automaker pointed out that the vehicle met all applicable government-mandated safety requirements and that it was as safe as other vehicles of its size and type. In fact, the Pinto did meet federal standards of the time. Standards did not require that cars should be able to withstand a rear-end collisions of a specified number of miles per hour until 1977. As Lacey commented, "Conforming to the letter of the law was hardly a defence for those executives inside Ford who discovered what a hazard to life the Pinto gas tank represented soon after the car was launched. . . . The question was how much that protection should cost."

Meanwhile, recalls followed. In October 1976 NHTSA (the National Highway Traffic Safety Administration) ordered 372,584 Pinto, Bobcat, and Mustang II models recalled because of potential fuel leaks from the engines. A few 1977 Pintos were recalled because they had rear-bumper nuts capable of puncturing the fuel tank. The 1976 Pintos were recalled because a 30-mph frontal impact could lead to fuel leakage.

The NHTSA announced in May 1978 that its own investigations "demonstrate that low-to-moderate-speed rear-end collisions of Pintos produce massive fuel tank leaks due to puncture or tearing of the fuel tank and separation of the filler pipe from the tank." In the "presence of external ignition sources," NHTSA determined, such fuel leaks could "result in fire." A dozen test rear-end collisions at 35 mph yielded two fires from fuel spillage. Moreover, government experts then knew of 38 cases of fuel-tank damage, leakage, and/or ensuing fires, which resulted in 26 fatalities.

Ford recalls 1.4 million Pintos

On June 15, 1978, Ford voluntarily recalled 1.4 million 1971–76 Pintos (the fuel tank had been redesigned beginning with the 1977 Pinto) and 30,000 similar-make 1975–76 Mercury Bobcats. Publicly insisting that Pin-

This owner warned other drivers to keep a safe distance behind her 1975 Ford Pinto.

tos were not defective, Ford agreed to make changes to "end public concern." The modifications cost Ford $40 million and included installation of two high-density polyethylene shields around the fuel tank.

Calling the shield installation "an inadequate technical fix; the cheapest way out," consumer activist Ralph Nader wanted Ford to retrofit the Pinto with the double-wall tanks used on bigger Fords. But adding a heavier, double-wall gas tank would add more weight, which would require a more powerful engine, a heavier transmission, bigger brakes, and so on. Retired Chrysler auto-safety director Roy Haeusler warned that many Pinto motorists were "sitting in virtual time bombs" due to unsafe tank designs.

Three teenagers die in Indiana crash

Then in August 1978 three teenaged girls driving a 1973 Pinto in Indiana were struck from the rear by a van. The Pinto burst into flames and all three girls died. Ford Motor Company was charged with reckless homicide, earning the automaker the dubious distinction of being the first corporation to be charged with the criminal offense. As the Indiana case opened, Ford was facing more than 50 civil suits and at least 2 class-action suits. Courts had already awarded plaintiffs as much as $6 million in damages. If found guilty, Ford faced a maximum $30,000 fine.

Indiana prosecutor Michael Cosentino chose not to try to prove that the design of the Pinto was flawed. Instead he argued a narrower issue: that Ford had allowed the Pinto "to remain on Indiana highways, knowing full well its defects." Ford pleaded not guilty. Ford's defense counsel was former Watergate prosecutor James F. Neal.

Indiana case did not set legal precedent

The point of law that attorneys for the plaintiffs pursued—that manufacturers of a product can be held criminally responsible for injuries resulting from the use of its product—would have set a precedent if the prosecution had won its case. But the court found Ford innocent on all charges. Judge Harold Staffeldt, who insisted that "a revolt against technology" was implicit in the case, refused to set legal precedent on the issue of criminal responsibility.

Though Ford was acquitted, it continued to suffer from the damaging publicity surrounding the Indiana case. The automaker eventually dropped the Pinto from its product lineup and replaced it in 1981 with the Ford Escort.

Impact

As Robert Lacey noted, "Scarcely a day seemed to pass without the name of Ford being linked to some damaging headline or other." Widespread public distrust of the American automotive industry resulted from the Pinto controversy and, in the mid-1960s, from the oversteering problem of the General Motors Corvair, which Ralph Nader had publicized.

Ford put dollars ahead of lives

The most damning accusation was that Ford put dollars ahead of lives. Mark Dowie declared in *Mother Jones* that Ford executives decided not to modify the Pinto after their "'cost-benefit analysis,' *which places a dollar value on human life,* said it wasn't profitable to make the changes" [Dowie's emphasis]. But cost-benefit analysis, which was an acceptable practice in the entire automotive industry, was federally sanctioned by the 1966 National Traffic and Motor Vehicle Safety Act. In fact, as Lacey reported, "the accountants at NHTSA came up with a price for a human life in the early 1970s—$200,725.00, the sum of twelve 'societal components.'"

Ford decided not to install the fuel valve in certain vehicles based on the NHTSA's own guidelines. Not installing the fuel valve would result in fires in 2,100 vehicles, they calculated, causing 180 deaths and 180 serious injuries. Based on the round figures of $200,000 per death, $67,000 per injury, and $700 per vehicle, Ford expected to pay $49.5 million in lawsuits if it did not modify the vehicle. In contrast, Ford claimed modifying the vehicles would cost $137 million, so by dispensing with the safety modification, Ford would save $87.5 million.

The automotive industry was finding out that safety and quality mattered to the public. "This was when making cars became more than a game," Lacey observed. "Quality and safety are just two among the wide range of concerns that keep a car executive busy—but they are quite predominant anxieties for the men and women whose names are on the letterhead. Outside directors only come into the hothouse atmosphere of headquarters one day a month, and for the other twenty-nine they have to live out in the ordinary world, . . . receiving angry letters from shareholders." If the board members had to be solicitous of the public, then so would the automakers. As Lacey direly noted, "If Detroit had shown the slightest willingness genuinely to police and regulate itself when it came to questions like crash-worthiness or the amount of lead in exhaust fumes in the early 1960s, it might have enjoyed a smoother entry into the consumer age."

Lacey pointed out: "The question is whether Ford and Iacocca exhibited all due care for their customers' safety when balanced in the complex carmaking equation that involves cost, time, marketability, and profit. Confronted with the facts of how Lee Iacocca and Ford actually built the Pinto, a number of juries across the United States concluded that they did not [exhibit due care]." The judicial system and Ralph Nader's "ugly revelations" about the automotive industry ultimately brought government regulation upon it and finally convinced the industry that it had to be responsive to its public.

Where to Learn More

Domestic Safety Defect Recall Campaigns. National Highway Traffic Safety Administration (1976 ed.): 21, (1977 ed.): 16, (1978 ed.): 12–14.

Dowie, Mark. "Pinto Madness." *Mother Jones* (September–October 1977): 18–32.

Lacey, Robert. *Ford: The Men and the Machine.* Boston: Little, Brown, 1986.

Panztur, Andy. "Pinto Criminal Trial of Ford Motor Co. Opens Up Broad Issues." *New York Times* (January 4, 1980): 1, 23.

"Pinto Ruling." *Time* (February 12, 1979): 87.

Rowand, Roger. "An All-New Ball Game for Engineers." *Automotive News* (January 21, 1980): 48.

Stuart, Reginald. "U.S. Agency Suggests Ford Pintos Have a Fuel System Safety Defect." *New York Times* (May 9, 1978): 22.

I V

Dams and Bridges

Austin Dam fails

Austin, Pennsylvania
September 30, 1911

Background

The afternoon of September 30, 1911, was sunny and "perfect" in the towns of Austin (population 3,200) and Costello (approximately 400 to 500 residents) in the mountainous country of north central Pennsylvania. In the 30-year-old town of Austin, townspeople were strolling down Rutherberger Street, enjoying the warm weather. Suddenly there was a loud blast—and seconds later a torrent of water roared through the valley, channeled by the valley's narrow, rugged walls.

The water's force burst underground gas mains, and after the wall of water abated, an errant flame ignited the leaking gas. Fire jumped from gas pipe to gas pipe—and it fed on whatever houses and buildings the flood had left standing.

Exaggerated initial estimates claimed that 1,000 people perished and 2,000 were homeless—later reports calculated the death toll from 50 to 149 people. The company that owned the dam—Bayless Pulp and Paper Mill—was the major employer for residents of Austin and the surrounding towns. Bayless built the concrete dam in 1909 to provide a storage reservoir for its water-intensive pulp and papermaking business. In January 1910, after intense winter rain and snowmelt, cracks appeared in the dam. The cracks were repaired, but no one suspected that serious foundation, design, or structural problems would be in the offing.

Poorly constructed concrete dam fails catastrophically, rupturing gas lines and precipitating the fire that destroys the industrial town of Austin, Pennsylvania.

Optimistic economic climate encourages Bayless to build bigger dam

Industry was booming in the early twentieth-century United States, and there was little government regulation to hamper it. In 1909 Bayless

An eyewitness to the Austin Dam disaster recalled: "The break ran across the front of the dam like the crack of a whip and the water gushed out with a force that carried a whole section . . . sixty feet away."

Pulp and Paper Mill decided to replace its small dam with a much larger one in order to expand production capability. No one guessed that, by altering the area's water resources, the project would endanger the entire town. Engineers designed the new dam and were to oversee its construction, but cost and time constraints brought pressure to bear to finish the dam's construction before the winter of 1909–10.

The town of Austin is in the valley of Freeman Run. The dam spanned this steep valley, 530 feet long at its base (across the floor of the valley) and 544 feet at its crest. The dam was 49 feet high at its crest (top) and 42 feet high at the crest of its emergency spillway. It measured 32 feet thick at its base and about 2.5 feet thick at crest. The reservoir, designed to contain 200 million gallons of water, filled one-half mile of the valley.

Concrete bonds boulder-sized rocks together

The dam built by Bayless was a gravity dam constructed with two major design components: concrete and earth. The dam's concrete consisted of 16,000 cubic yards of cyclopean concrete, a cement substance that bonded together boulder-sized ("cyclopean") pieces of native quarried sandstone rock. Cyclopean concrete differs markedly from the gravel-sized rock used as aggregate in late-twentieth century concrete construction practice.

The concrete for the Bayless reservoir was placed to break joints in stepped sections that were dovetailed together. The concrete was also reinforced with twisted steel rods 1.25 inches in diameter and 25 feet long, placed 2.7 feet apart from each other. Steel-reinforced pilasters, or buttressing sidewalls, were placed on either side of the spillway. Extra reinforcement was installed in the top, or crest, of the dam.

The second major component of the dam was an earthfill embankment that was rolled into place against the concrete on the upstream, or reservoir, face of the concrete dam. This embankment was 27 feet high, and it sloped at 3 feet horizontally for every 1 foot in vertical height down from the point where its top met the concrete dam into the floor of the reservoir.

The area's bedrock is shale and sandstone

The dam's foundation was composed of the rock in the valley floor: thin layers of shale nestled between horizontal layers of sandstone. The builders excavated about 8,000 cubic yards of soil. Then at the base of the dam they compacted and cemented alluvium, or very small gravel, together with loose rock from the area. They drilled holes 8 feet into bedrock to explore for fractures and fissures.

Construction crews prepared the surface for the base of the dam by washing and grouting it with cement. Then they founded the dam on concrete footing, which extended 4 feet into bedrock and was 4 feet wide. They set the abutments, or sides, 20 feet into bedrock. No records exist to show whether or not they conducted studies of the permeability of the rock layers or to determine the necessity of a cutoff wall, which would prevent seepage from traveling through the horizontal sandstone under the dam.

Two cracks occur: Concrete contraction is blamed

The dam was still under construction as the winter of 1909–10 approached. Temperatures were below freezing when crews placed some of the concrete, and they hurriedly finished the final stages of construction. When the dam was completed about December 1, 1909, there was already a visible crack running from the crest of the dam vertically to the ground. This crack actually passed through—not just around—the cyclopean sandstone boulders. By the end of that month a second crack appeared. As there was no measurable settlement of the dam and no water in the reservoir yet, officials concluded that both cracks resulted from contraction of the concrete. By six weeks later they had filled the reservoir, placing the maximum load on the fresh concrete.

On January 17, 1910, a warm spell and heavy rains caused rapid snowmelt. Four days later flood water was pouring over the spillway. The next day heavy seepage emerged from the toe of the dam, and cracks also appeared near the bottom of the downstream face of the dam. There was too much water in the reservoir, and Bayless authorities knew they had to lower the operating level of the reservoir. They authorized the use of dynamite to blast two notches through the crest. After the water level came down, workers made hasty, superficial repairs because of the time of the year. The two notches were not filled in—and the reservoir allowed to rise again—until late September 1911. The ineffectiveness of these repairs would become all too obvious on the last day of the month.

Details of the Failure

On that September afternoon, a hole opened between the dam and the west abutment, and it continued to spread. The force of the water shifted major portions of the dam. Pieces overturned, and a large section of the center of the dam moved downstream—as if on rollers. No longer held in by the wall, the water exploded into Austin. Because the valley walls were high and nearly vertical, the water could not spread out. An eyewitness saw leakage emerging from five or six cracks on the face of the dam as it came apart. "One after another six great masses of concrete came crashing through the valley just as though you were to knock down one domino after another," he exclaimed. "The break ran across the front of the dam like the crack of a whip and the water gushed out with a force that carried a whole section . . . sixty feet away."

A wall of rushing water with white foam on its crest

Someone rang a fire bell as a warning, and people ran into the streets to face the crushing flood. Those who escaped into the surrounding hills

Austin, Pennsylvania, in the aftermath of the 1911 failure of the Austin Dam. Flooding and the ensuing gas-main fire claimed from 50 to 149 lives.

later testified to the roar of the rushing water and the white foam on its crest. Homes were ripped from their foundations and capsized into the torrent. Large machines at the Buffalo & Susquehanna Railroad were overturned onto their operators.

The largest gas main in the business district of Austin ruptured and burst into flame. Sucked by the draft in the wake of the flood, the fire traveled from house to house. Other pipes broke and fed the fire. The stove wood that was waiting shipment at Standard Lumber Company added more kindling. Bayless Pulp and Paper Mill—damaged by flood—was now ravaged by fire.

Austin's fire protection, limited at best, was wiped out by the flood. The wall of water left injured and dead in the streets, and the raging fire prevented rescuers from reaching them. State senator Frank E. Baldwin was among the injured, but his wife, children, parents, and sister were among the dead. Three miles downstream of Freeman Run was the town of Costello, and it was completely leveled by the flood. Eight miles below

Costello was the town of Wharton, which suffered heavy damage but no fatalities.

Telephone lineman calls in medical help

A telephone company lineman saw the flood strike the town and called for help from a home on the hillside. An hour later a train with doctors, nurses, food, and medical supplies was bound for Austin. Assistance also came by automobile over the mountain roads. The hospital, sited on the hillside, escaped the devastation. So did a schoolhouse, but every church in town was destroyed. The fire burned itself out by midnight, when the rescue effort could begin in earnest. Stories of heroism and heartbreak began to emerge.

On October 1, 1911, the Pennsylvania State Water Commission and the District Attorney of Potter County began two separate inquiries into the dam's failure. Chalkley Hatton, the engineer who designed the dam and reviewed its condition after the 1910 failure, maintained that his original design was sound.

The engineer pointed out, however, that the foundation had been undermined by seepage in 1910. Hatton said he notified Bayless Pulp and Paper Company at that time that they should construct a concrete cutoff trench upstream of the dam and reinforce the downstream face of the dam. Hatton stated that the repairs he recommended were never made.

The engineer explained that foundation exploration had been conducted during the original design and construction, but it gave no hint that seepage might ever be a problem. Hatton also blamed the dam's problems on the placement of concrete during freezing weather. He further argued that there was inadequate time allowed for the concrete to cure before the reservoir was filled.

Actual water pressure was much higher than estimated

When the failure analysis reports came in, they revealed that the cracks where the concrete failed, called break lines, corresponded to work done on different days. This evidence supports the theory that hurried construction contributed to the failure. In addition, the faces of all but two of the cracks were discolored, which meant that the September 30 failure occurred on older break lines.

Excavations into the failed foundation proved that seepage of water on the upstream side was a major factor in the dam's failure. Rock had slid from 6 feet below the concrete foundation to about 12 feet upstream of the base of the dam. When the hydrostatic pressure uplifting the dam was recalculated based on these post-failure observations, it was found to be much higher than that estimated during design.

Present researchers agree with investigators at the time that all technical aspects of Austin Dam were poor—even by 1909 standards, and certainly with modern ones. Foundation conditions were not studied adequately, and ineffective foundation treatment was applied. Austin Dam is a gravity-type dam. A gravity-type dam uses mass to resist motion, and the mass—including the area at the base of the dam, the foundation embedment, and the other methods of anchorage—was insufficient.

The construction failures were obvious and numerous: they included use of weak, oversized aggregate placed in improperly cured concrete in freezing weather. The appearance of the first crack proclaimed that the dam had already failed, but this fact was ignored and construction continued. The appearance of the second crack repeated the ominous message—and was ignored again.

When more cracks and seepage from the toe of the dam occurred in January 1910, they declared the failure of the entire structure as well as the founding bedrock, yet these signs were conspicuously ignored. Blasting holes in the dam, which they did to bring down the water level, was an astounding, absurd blunder that probably weakened whatever remained of the structure's integrity. When the dam's owner and operator, Bayless Pulp and Paper Mill, disregarded the engineer's recommended repairs, the dam's fate was sealed. Although the reservoir should never have been filled after cracks were observed in December 1909, it definitely should have been abandoned completely after the near disaster of January 1910. Inexplicably, even these warnings were ignored.

Impact

The failure of Austin Dam was costly in fatalities, injuries, and loss of private property. This failure and other, similar failures of structures owned by industries and individuals eventually resulted in state and federal legislation for responsible design, operation, and maintenance of the hundreds of such dams nationwide.

Alexander Rice McKim, inspector of dams and docks for the Conservation Commission of the State of New York, visited Austin, Pennsylvania, and also inspected the dams in New York State. He declared: "I find the dams in this State generally in a very neglected condition, even when originally well constructed. The responsibility and the danger of impounding waters behind insufficient barriers is generally not recognized." With the benefit of hindsight, the modern researcher can see every event in Austin Dam's history clearly: multiple errors were compounded, and there were multiple opportunities to avert disaster, but they were never heeded.

Where to Learn More

"Austin Dam Failure." *Engineering Magazine* (November 1911): 253–257.

"Austin Dam Was Weak." *New York Times* (December 8, 1911): 10.

"Failure of the Austin Dam." *Scientific American* (October 14, 1911): 331+.

Jansen, Robert B. "Austin Dam." *Dams and Public Safety: A Water Resources Technical Publication.* Washington, DC: U.S. Department of the Interior, Water and Power Resources Service, 1980, 119–120.

"Nearly 1,000 Dead in Austin, Pa., Devastated by Flood and Fire." *New York Times* (October 1, 1911): 1.

"Two Inquiries Begun into Dam's Breaking." *New York Times* (October 2, 1911): 3.

Vaiont Dam landslide

Near Longarone, Italy
October 9, 1963

Background

On the night of Wednesday, October 9, 1963, the residents of Longarone, with a population of 4,000, and several other villages in the steep-sided Piave River Valley heard a roar like thunder. Fifteen minutes later a shock wave preceded a 300-foot-high wall of water, rock, mud, and debris that swept through the valley. Three-fourths of the population perished, and all traces of the many fourteenth-century buildings vanished. All that was left was a wasteland of mud. Initially the stunned world thought that the dam must have failed. The Vaiont Dam is the highest (869 feet) thin-arch dam in the world. When daylight broke, the dam—still standing—had miraculously sustained only minor damage. The flood occurred because a huge landslide from neighboring Mount Toc filled the reservoir (the lake behind the dam). The water then surged into the valley below.

Thousands die when a huge landslide fills the reservoir behind the Vaiont Dam, forcing a wall of water over the dam. The flood destroys villages in the valley below.

No electricity in the Piave River Valley until the 1960s

Electricity was not available in the villages of northeastern Italy until the early 1960s. This was when a series of water collection and power systems were built in the picturesque river valleys of the Alps, where Italy, Switzerland, and Yugoslavia meet. The Piave River and its tributary, the Vaiont, flow toward the Adriatic Sea, which the Società Adriatica di Eletricitta (SADE) felt would be a good site for electrical generation. They appointed Carlo Semenza, manager of SADE's hydroelectric department, to supervise the Vaiont project.

The site for the dam was a narrow, steep-walled gorge. The dam itself was designed as a double-curvature arch (curving both from top to bot-

The Vaiont Dam before a landslide filled the reservoir. A wall of water then surged over the dam, destroying villages in the valley below.

tom and side to side), 625 feet long, 11 feet thick at the crest (top), and 73 feet thick at the base. A project of gargantuan proportions, it would require the building of other facilities as well: an overflow spillway for emergencies, two outlet tunnels through the rock wall of the gorge (also for use in releasing water), power tunnels for hydroelectric power, the underground power plant at the downstream base of the dam, and an office and hotel high above the dam on opposite rock walls for the workers. Construction began in 1956 and would not be completed until 1960.

Studies find bedrock fractured, weak, and pitted

Extensive analyses were performed at the site of the proposed dam. A scale model was analyzed at the Istituto Sperimentale Modelli e Strutture (ISMES), a famous structural laboratory located in Bergamo, Italy. Another scale model was built after the foundation had been excavated; this would allow exact analysis of the support provided by the limestone bedrock under the dam and up the abutments (sides where the dam and rock meet).

When the foundation for the dam was excavated, the rock was found to be fractured, weak, and pitted with solution cavities (holes dissolved in the limestone by water). To strengthen the foundation and abutments, tunnels were drilled into the rock, and grout (cement) was injected through the tunnels to fill these gaps. This procedure is commonly done to dam foundations to improve the natural conditions. An unusual addition to the site was the drilling of a tunnel under part of the reservoir from upstream of the dam to over a mile further upstream. The purpose of this bypass tunnel was to allow water to flow to the dam if landslides should ever block the lower (near dam) part of the reservoir. This "unique and remarkable" tunnel was built in 1961, after the dam was completed. A landslide had flowed into part of the reservoir in 1960, and designers wanted to preserve the capability of the dam to release water for generating power and also for emergencies.

Area geology is described as "adverse"

The reservoir area was also studied during development of the project, with special attention paid to the stability of Mount Toc. Major landslides were known to have occurred in the Vaiont valley over geologic time. The geology was described as "adverse," meaning that the materials and the natural shape or geometry of the area were a bad combination. Rock in the area had formed from thick sediments, including limestone, claystone, and marl. These were low or weak in shear strength and were badly affected by ground water, which caused natural joints and bedding planes to enlarge and weaken, creating solution cavities.

During construction of the dam, landslides occurred and cracks appeared on the mountain slopes. These were repaired and stabilized with concrete, but instruments installed in the mountainside showed additional stresses and strains, which indicated potential for movement. Slow creep of soil and rock down the face was also observed, accompanied by small earth tremors, like local earthquakes, from 1960 through 1963. This movement—measured to be as many as 12 inches per week between 1960 and 1961—decreased thereafter until the autumn of 1963.

When a landslide occurred on November 4, 1960, project managers decided to build the bypass tunnel. When the bypass tunnel was drilled in 1961, neither the small-diameter exploration holes nor the actual tunnel surfaces revealed indications of significant slide planes (failure surfaces that precipitate landslides). Cracks continued to appear on the mountain's slopes and to grow above the 1960 slide, including the outline of the massive slide that would occur in 1963. The reservoir was left empty for two years because of concern over these observed changes. From 1960 to April 1963, the lake level was also kept low as a precaution. Creation of the lake, however, added new and weakening water forces. Heavy rains in August and September 1963 caused water pressure and the weight of the rock to increase, while the strength of the rock decreased. The rate of slope creep began to increase again on September 18, 1963.

Details of the Disaster

Intense rainfall occurred again September 28 through October 9. On Mount Toc grazing animals abandoned the area of concern. The mayors of Erto and Casso, villages on the shore of the reservoir and in the landslide area, cautioned fishermen that dangerous waves might be caused by landslides. Erto was evacuated. The level of the reservoir was lowered 10 feet between September 28 and October 8. On October 8 technicians who were monitoring instruments realized that the huge rock mass was moving—and that it was five times larger than previously estimated. Reservoir operators lowered the lake level further, but heavy rains canceled out their efforts.

Mount Toc dumps 300 million cubic yards of rock into reservoir

At 10:39 P.M. on October 9, 1963, the rock mass estimated at over 300 million cubic yards crashed 1.2 miles down the mountainside at 100 feet per second. Geologists in Vienna, Austria, and Brussels, Belgium, recorded the landslide as a strong earth tremor. It set off a shock wave that blasted water and rocks almost 800 feet into the air. As the landslide toppled into the reservoir, the displaced water reared up 330 feet above the rim of the dam. It ripped through the village of Casso, high on the slope above the dam. It demolished the workers' hotel, then surged to the other abutment and swept away the office for the dam's operations. No one survived.

One mile downstream from the dam, where the Vaiont River enters the Piave River, the town of Longarone was struck at 10:43 P.M. The raging waters were 230 feet high. The villages of Pirago, Rivalta, Villanova, and Fae fell in the flood's wake, as did the smaller hamlets of Codissago and Castellavazzo. Within 16 minutes—by 10:55 P.M.—the wall of water was a mere flood downstream, and the screaming waters gave way to a deadly calm. In Pirago only a church tower and some electric power poles remained standing. Railroad tracks were twisted into spirals, massive steel doors at the dam were ripped away, all means of communications were destroyed, and access to the valley was blocked by mud, rubble, and dead cattle.

Dam survives landslide, is hardly damaged

When daylight broke, the scene was a mud-drenched holocaust. The dam, curiously beautiful, survived almost untouched, with only a small notch cut out of the crest. The colliding water and rock is estimated to have been about 4 million tons of force, many times greater than the dam's design pressure. The blast of air and water opened and destroyed all of the internal workings of the dam and the underground power plant. Behind the dam, two-thirds of the reservoir was filled with a new mountain that towered 574 feet above the water level. And downstream from the dam, only an occasional building could be seen in the quagmire of mud.

Rescuers had difficulty reaching the scene. The swollen river was twice its normal size. Entire villages were buried in a sea of mud. Bodies were tossed into treetops. There were so many dead and so few survivors, identification of bodies became impossible. Even the final death toll is vague. No one knows whether all the bodies were recovered, and, in any event, official population records in each hamlet were lost. In at least one miracle, two children were found alive in the cellar under their buried home on October 12, three days after the event.

Ministry of Public Works criticizes choice of site

Official inquiries into the disaster were highly charged with emotions and political pressures. Two types of inquiry were mandated by the Italian government. A court of inquiry was established quickly and was under pressure from the Communist party of the Italian Parliament. The engineers through the Superior Council of Italy's Ministry of Public Works also studied the disaster. Italy's nationalized electrical industry, including SADE, was found to be at fault for allowing the site to be used, and the project's builders and engineers were tried in the courts as having responsibility for the tragedy. Inquiries determined that the suitability of the location of the dam and reservoir was still being researched at the time of the disaster. The problems were several and were mostly related to the reservoir and general site location, not the dam itself.

The hazards of landslides, creep, and tremors due to stresses in the mountainside were well-known, but actions were taken neither to stop the project nor to protect the population. Exploratory holes had been drilled in the reservoir and landslide area, but none was deep enough to intercept the slide plane of the massive landslide. In fact, the monstrous size of the rock mass had never been approximated or even guessed, despite the pattern of cracks visible on the mountainside and the data gathered from instruments, local geology, and other indicators. The lack of an early warning system to protect the local population was also a major contributor to the death toll.

Some thought was given initially to clearing the reservoir, but the Superior Council concluded that the task was impossible. In addition to the mass of earth to be removed, there were questions about where to place the huge volume of earth as well as how to pay for such a project. Instead they eliminated Vaiont Dam as a power source within the area's hydroelectric network. Then villages downstream were rebuilt and new factories were constructed on the valley floor.

Impact

The most horrific impact of the landslide at Vaiont Dam was the loss of 3,000 lives. The disaster was not foreseen and may have been unforesee-able, because this set of compounding conditions had never occurred before. Other dam projects have been considered failures because reservoirs filled with silt or were unable to hold water. Vaiont was the first case in which a landslide filled a reservoir so completely, let alone with such devastating consequences. That the dam itself survived is a triumph of engineering, although that fact is dwarfed by the immensity of the disaster.

Preconstruction studies must now analyze how site will behave after reservoir is filled

In fairness to the Vaiont engineers, no landslide had previously been known to move so quickly over such a long distance. Also, it may not have been possible to envision movement of a mass of that scale, even though indications seem clear in hindsight; this topic is still debated by engineers. For today's projects, considerable study is given to the suitability of a site for both the dam and reservoir; much like a bathtub, all sides must be strong and stable if it is to hold water successfully. The Vaiont disaster showed that the geology of the area must be studied thoroughly and under two conditions: as it exists before the dam is constructed and as it will behave after the reservoir is influenced by the weight, volume, and movement of water.

The other lesson from the Vaiont disaster was the need for protecting the public. The mayors of two of the villages above the dam had the foresight to warn and evacuate their citizens, but no word of caution was given by any person, official, or agency to the towns downstream. Early warning systems that allow enough time for evacuation are now required practice. Such systems allow changing and compounding circumstances to be considered when warnings are issued (for example, the rainfall and slope movement data from the Vaiont site), and they require actions from a series of agencies such as police and fire departments, so that if one element of the early warning system fails, others are available.

Where to Learn More

"Alpine Dam Resists Slide—U.S. Army Aiding Rescuers." *New York Times* (October 11, 1963): 1.

"Alps Spawn Notable Dams." *Engineering News-Record* (November 2, 1961): 32–40.

"Before and After at Vaiont." *Engineering News-Record* (December 5, 1963).

Canning, John, ed. *Great Disasters: Catastrophes of the Twentieth Century.* Stamford, CT: Longmeadow Press, 1976, 76–77.

Frank, Beryl. *Great Disasters of the World.* Austin, TX: Galahad Books, 1981, 160–163.

"Giant Dam Falls in Italy; Hundreds Reported Dead as Water Engulfs Town." *New York Times* (October 10, 1963): 1.

Jansen, Robert B. *Dams and Public Safety: A Water Resources Technical Publication.* Washington, DC: U.S. Department of the Interior, Water and Power Resources Service, 1980, 214–220.

Kiersch, George A. "Vaiont Reservoir Disaster: Geologic Causes of Tremendous Landslide Accompanied by Destructive Flood Wave." *Civil Engineering* (March 1964): 32–39.

Legget, Robert F., and Paul F. Karrow. *Handbook of Geology in Civil Engineering.* New York: McGraw-Hill, 1983.

"Lessons from Vaiont." *Engineering News-Record* (October 24, 1963): 84.

"Vaiont a Total Loss?" *Engineering News-Record* (October 24, 1963): 21.

"Vaiont Dam Survives Immense Overtopping." *Engineering News-Record* (October 17, 1963): 22–23.

"Vaiont Dropped as Power Source." *Engineering News-Record* (October 31, 1963): 18.

"Vaiont Slide—Seven Years Later." *Civil Engineering* (June 1970): 86.

Tacoma Narrows Bridge collapses

Puget Sound, Washington
November 7, 1940

Background

Although the newspapers were filled with international stories on Saturday, November 8, 1940—England's Royal Air Force was bombing Berlin, the Luftwaffe was bombing London, and Franklin D. Roosevelt was returning to Washington, D.C., for an unprecedented third presidential term—the *New York Times* placed a single-column national story on its front page: "Big Tacoma Bridge Crashes 190 Feet into Puget Sound."

Third-largest suspension bridge fails at five months old

The Tacoma Narrows Bridge was the third largest suspension span bridge in the world and only five months old. The center span, measuring 2,800 feet, stretched between two towers 425 feet high. The side spans were each 1,100 feet long. The suspension cables hung from the towers and were anchored 1,000 feet back towards the river banks. The designer, Leon Moisseiff, was one of the world's foremost bridge engineers. He and his partner, Fred Lienhard, were well known for their calculations for determining load and wind forces used by bridge designers everywhere.

During the 1930s the widespread design trend was toward "streamlining" every product from teakettles to locomotives to airplanes. Following this trend, Moisseiff intended to produce a very slender deck span (roadway of the bridge) arching gently between the tall towers. His design combined the principles of cable suspension with a girder design of steel plate stiffeners—running along the side of the roadway—that had been streamlined to only 8 feet deep.

The failure of the five-month-old bridge ends the search for "a slender ribbon bridge deck" and creates a new academic specialty, wind dynamics engineering.

For months, the motions, while disturbing, had been symmetrical, and the roadway had remained flat. The twisting motions fatigued the metal and caused the collapse.

Bridge is nicknamed Galloping Gertie

The $6.4 million bridge opened with much fanfare on July 1, 1940. It was celebrated as a defense measure to connect Seattle and Tacoma with the Puget Sound Navy Yard at Bremerton, Washington. Owned by the Washington State Toll Bridge Authority, the bridge was financed by a Public Works Administration grant and a loan from the Reconstruction Finance Corporation. Construction was completed in only 19 months.

The bridge gained notoriety even before it opened—people who experienced its strange behavior nicknamed it Galloping Gertie. Workers complained of seasickness from the pitching and rolling of the deck caused by undulations. After the bridge opened to traffic, it became a challenging sporting event for motorists to cross even during light winds, and complaints about seasickness became common.

Narrows bridge is three times more flexible than the Golden Gate

Despite Moisseiff's reputation as a top-ranked engineering consultant, State and Toll Bridge Authority engineers were more than a little nervous about the behavior of the slender, two-lane span just 39 feet wide. Its shallow depth in relation to the length of the span (8 feet to 2,800 feet) resulted in a ratio of 1:350, nearly three times more flexible than the Golden Gate or George Washington bridges. Engineers tried several methods to stabilize, or dampen, the oscillations—the up-and-down waves of the deck.

The first method involved attaching heavy cables—called tie-down cables—from the girders and anchoring them with 50-ton concrete blocks on shore. The cables soon snapped, and another set installed in a second try lasted only until the early morning hours of November 7. A more successful method, a pair of inclined stay cables connecting the main suspension cables to the deck at mid-span, remained in place but proved ineffective. Engineers also installed a dynamic damper, a mechanism consisting of a piston in a cylinder, which also proved futile: its seals were broken when the bridge was sandblasted prior to being painted.

Movies monitor bridge oscillations and record its collapse

Measurements and movie camera films taken over several months gave the engineers a good idea of how the bridge was moving in the wind. Charting the oscillations and vibrations, they discovered the movements were peculiar. Rather than damping off (dying out) very quickly as they did in the Golden Gate and George Washington bridges, Galloping Gertie's vibrations seemed almost continuous. The puzzle was that only certain wind speeds would set off the vibration—yet there was no correlation between the vibration and the wind speeds.

Led by Frederick B. Farquharson, a professor at the University of Washington engineering school, the study team applied actual measurements of the bridge movements to a scale model, hoping to find ways to stabilize it. They suggested installing additional stabilizing cables, attaching curved wind deflectors, and drilling holes in the girders to permit wind to pass through. Their report came out just one week before the bridge collapsed.

Interest in the phenomenon rose with the onset of the brisk fall winds pushing through the valleylike narrows that lie between the cities of

Tacoma and Bremerton. The public and the engineers kept watch, and when the bridge snapped, it became one of history's most documented disasters. Motion picture and still cameras recorded the collapse on film. A newspaper reporter was a mid-span survivor, supplying a firsthand account of the disaster.

Details of the Collapse

Witnesses included Kenneth Arkin, chairman of the Toll Bridge Authority, and Professor Farquharson. As Arkin later recalled, that morning he had driven to the bridge at 7:30 to check the wind velocity, and by 10:00 he saw that it had risen from 38 mph (miles per hour) to 42 mph. The deck rose and fell 3 feet 38 times in one minute. He and Farquharson halted traffic and watched while the bridge waved up and down and began to sway from side to side. Then it started twisting.

Insurance billboard boasts, "As Safe as the Narrows Bridge"

Meanwhile, Leonard Coatsworth, the newspaper reporter, abandoned his car in the middle of the bridge because the undulations made further driving impossible. He tried to retrieve his daughter's pet dog from the car, but he was thrown to his hands and knees. Other reporters watched him, "hands and knees bloody and bruised," as he crawled 500 yards while the bridge pitched at 45 degree angles and concrete chunks fell "like popcorn." The driver and passenger of a logging truck told a similar tale of jumping to the deck and crawling to one of the towers, where they were helped by workers.

By 10:30 the amplitude (the distance from crest to valley) of the undulations was 25 feet deep. The suspender ropes began to tear, breaking the deck and hurling Coatsworth's car and the truck into the water. When the stiffening girder fell 190 feet into Puget Sound, it splashed a plume of water 100 feet into the air. Within 30 minutes the rest of the deck fell section by section, until only the towers remained, leaning about 12 feet toward each shore. Overlooking the bridge was an insurance company billboard that bragged, "As Safe as the Narrows Bridge." The slogan was covered up before the end of the day. The only casualty of the bridge's collapse was the dog.

Moisseiff's first public comment was, "I'm completely at a loss to explain the collapse."

The center span of the Tacoma Narrows Bridge writhes under a gale force wind. Moments later the bridge collapsed.

Chief engineer claims his designs were changed

Charles E. Andrew, chief engineer in charge of construction, said that the collapse was "probably due to the fact that flat, solid girders were used along the sides of the span." He pointed out that his original plans called for open girders but said "another engineer changed them." He compared Galloping Gertie with New York's Whitestone Bridge, which had been completed a year earlier. It was the only other large bridge designed with web-girder stiffening trusses, "and these caused the bridge to flutter, more or less as a leaf does, in the wind. That set up a vibration that built up until failure occurred."

The Whitestone, now known as the Bronx-Whitestone, was indeed very similar to the Tacoma Narrows. However, the stiffening trusses were twice as heavy and the deck twice as wide. Its designer, Othmar Ammann, had consulted Moisseiff, as had the designers of such structures as the Golden Gate and the San Francisco–Oakland Bay bridges. The Whitestone is 4,000 feet long, with a 2,300-foot main span and towers 375-feet high. It, too, had early oscillation problems, but they had been successfully damped with a device similar to the one tried on Galloping Gertie.

The deck was too slender

After the Tacoma bridge failed, Ammann realized the danger of the too-slender deck, and he insisted that a new steel truss be superimposed onto the deck of the Whitestone. His redesign, which also added cable stays between the towers and deck to prevent twisting, stabilized the bridge. In 1990, a tuned mass damper (a fairly recent invention) was added to the deck during a rehabilitation.

The first investigations into the collapse of Tacoma Narrows detailed how the bridge had come apart. For months the motions, while disturbing, had been symmetrical, and the roadway had remained flat. The lampposts on the sidewalks stayed in the vertical plane of the suspension cables even as they rose, fell, and twisted. But on November 7 a cable band slipped out of place at mid-span, and the motions became asymmetrical, like an airplane banking in different directions. The twisting caused metal fatigue, and the hangers broke like paper clips that have been bent too often.

Impact

Why did Galloping Gertie twist so violently when other bridges survived gale-force winds? Engineers looking into the problem explained that winds do not hit the bridge at the same angle, with the same intensity, all the time. For instance, wind coming from below lifts one edge, which pushes down the opposite. The deck, trying to straighten itself, twists back. Repeated twists grow in amplitude, causing the bridge to oscillate in different directions.

Wind engineering becomes new academic specialty

The study of wind behavior grew into an entire engineering discipline called aerodynamics, parallel to its study in the airplane industry. *Vortex shedding* and *flutter* were added to the vocabulary. A vortex is a spiral that

This new bridge was opened in 1950. Its four-lane deck and stiffening trusses form a box design that resists torsional forces. Excitation is controlled by hydraulic dampers at the towers and at mid-span.

can be seen in the wake of a ship or, in a wind tunnel, by wisps of smoke added to study wind behavior. Some vortices do not affect oscillation, but others form a flutter-like pattern that has the same frequency as the oscillating bridge. Bridges, buildings, and other exposed structures are now designed by first testing models in a wind tunnel. With the development of graphic capabilities, some of this testing is now done on computers.

Dozens of papers are published each year about these subjects; nevertheless, misconceptions continue. K. Yusuf Billah of Princeton University noticed in 1990 that even physics textbooks were citing the wrong cause for the collapse of the Tacoma Narrows Bridge. He and Robert H. Scanlan of Johns Hopkins University "set the record straight" in the February 1991 issue of the *American Journal of Physics*. The textbooks were claiming "forced resonance (periodic natural vortex shedding)" as the cause. Billah and Scanlan pointed out, "The aerodynamically induced condition of self-excitation [vibration] was an interactive one, fundamentally different from forced resonance."

Narrows bridge is scrapped

Two years after the collapse, the remains of the Tacoma Narrows Bridge were scrapped. In 1950 the state opened a new $18 million bridge designed by Charles E. Andrew and tested in wind tunnels by Farquharson and other engineers. Four lanes, a deck 60 feet wide, and stiffening trusses 25 feet deep form a box design that resists torsional forces. Excitation is controlled by hydraulic dampers at the towers and at mid-span.

The Tacoma Narrows Bridge collapse remains one of the most spectacular failures in the history of engineering. It is certainly one of the best known because of the movie film and widely reproduced still photos, but as Billah and Scanlan demonstrated, the aerodynamic details tend to become blurred. The new science of wind engineering, however, is now routinely applied to every type of structure.

Where to Learn More

Alexander, Delroy. "A Lesson Well Learnt." *Construction Today* (November 1990): 46.

Ammann, O. H., T. von Karman, and G. B. Woodruff. "The Failure of the Tacoma Narrows Bridge." *Report to the Federal Works Agency.* Washington, DC: Federal Works Agency, March 28, 1941.

"Big Tacoma Bridge Crashes 190 Feet into Puget Sound." *New York Times* (November 8, 1940): 1.

Billah, K. Yusuf, and Robert H. Scanlan. "Resonance, Tacoma Narrows Bridge Failure, and Undergraduate Physics Textbooks." *American Journal of Physics* (February 1991): 118–124.

Farquharson, Frederick B. "Collapse of the Tacoma Narrows Bridge." *Scientific Monthly* (December 1940): 574–578.

Goller, Robert R. "Legacy of Galloping Gertie 25 Years After." *Civil Engineering* (October 1965): 50–53.

Jackson, Donald C. *Great American Bridges and Dams.* Walnut Creek, CA: Preservation Press, 1988, 327–378.

Peterson, Ivars. "Rock and Roll Bridge." *Science News* (1990): 244–246.

Petroski, Henry. "Still Twisting." *American Scientist* (September–October 1991): 398–401.

Zilwaukee Bridge fails

Saginaw, Michigan
Late August 1982

A segment of a concrete-span freeway bridge under construction nearly collapses, delaying its completion for years and adding millions of dollars to the project's cost.

Background

The near-collapse of a portion of the Zilwaukee Bridge during its construction caused no fatalities or injuries—but it generated substantial controversy for its enormous cost and for the unusual techniques applied to repair it. The new freeway overpass was designed to replace a drawbridge over the Saginaw River—a perennial bottleneck for traffic on I-75 near Michigan's thumb, at the gateway to Michigan's northern recreational areas. The Zilwaukee Bridge was supposed to be completed in 1983 for a total cost of $76.8 million.

Then a segment of the bridge suddenly shifted beneath a load imposed by construction equipment in August 1982. State officials and contractors began a lengthy dispute about who would assume financial responsibility for repairing the damage. Painstaking repairs brought the price tag to nearly $133 million, and the bridge finally opened to traffic in 1988, five years later than originally scheduled.

Cantilevered bridge built from precast girders

During its construction, a cantilevered bridge is supported only on one vertical pier. The horizontal section projects outward toward the other pier until they are joined in the middle. One and one-half miles long, the Zilwaukee Bridge is a cantilevered bridge consisting of two parallel structures—one for the northbound lanes, one for the southbound lanes. It was assembled from 1,592 precast concrete box girders. Concrete piers divide the bridge into 25 spans on the northbound side and 26 southbound. Each girder averaged 73.5 feet wide by 8 feet long and weighed roughly 140 tons. They were attached one by one to cantilevered sections of the existing structure. Delivered by truck across previously completed spans, the girders were lowered into position by a steel truss launching girder, which held them in place while grout and temporary prestressing were applied. The launching girder itself spanned unfinished sections of the bridge, while it rested upon cantilevered segments.

Details of the Accident

Loads imposed on cantilevers differ greatly during and after construction of a bridge. During bridge construction, while cantilevers are anchored at one end and not at the other, the loads imposed are often far greater than loads imposed after completion. To prevent mishaps, the movement of tons of material and equipment must be carried out with precision, and the anchored side of the cantilevers must be strong enough to handle the massive loads.

Section of new bridge sags, twists, and drops

On a Friday night in late August 1982, a crane operator working on a section of the bridge over pier North-11 had just lifted a 157-ton box girder from a truck. The section began to sag, twisting to the side and deflecting 5 feet downward at its free end, where normal deflection should have been only 6 inches. The other end of the section, which was attached to an expansion joint on the other side of the pier, rose about 3 feet. A temporary 35-foot spacer block positioned within the joint to immobilize it cracked under the load, which freed the section to begin moving.

Predictably, initial concern centered around the possibility that the entire section would collapse. A state bridge design engineer, Adrianus Van Kampen, addressed the issue, stating that "if the prestressed cantilever can be returned to position, construction can continue. There is no danger that it will collapse; our figures show it is stable." However, not everyone was as optimistic.

New damage raises disputes on all sides

The Michigan Department of Transportation (MDOT) quickly sent experts from its research laboratory to assess the damage, and they found that the movement of the superstructure had two major consequences. First, a section of the footing supporting the pier had sheared nearly com-

An engineering professor investigating the mishap declared that workmanship, which had been adequate at the start of the project, had deteriorated in the area where the accident occurred.

pletely through. This caused the pier to shift slightly out of plumb. Second, the movement of the superstructure had damaged the bearings upon which it rested at the top of the dual 104-foot pier columns.

The project was a troubled one and was already more than two months behind schedule when the accident occurred. Not long after the mishap, accusations about its cause were being hurled by several parties. Involved in the dispute were the bridge designers, BVN/STS Consulting Engineers of Indianapolis, Indiana; the builders, Stevin Construction of the Netherlands in joint venture with Walter Toebe Construction of Wixom, Michigan; and MDOT, the overall administrators.

Engineers claim MDOT supplied faulty specs

Some time prior to the mishap, Stevin engineers found that they had to double the weight of the launching girder in order to manipulate the precast box girder sections. The new launcher weighed 1,000 tons, and Stevin claimed that its original specification had been based on faulty calculations supplied by MDOT. Early speculation therefore pinned the accident on the weight of the launcher.

Teams of investigators swarmed to the site. MDOT brought in Howard, Needles, Tammen & Bergendoff (HNTB) of Kansas City, Missouri, to report on the cause and to design repairs. The joint venture contractor retained both T. Y. Lin International of San Francisco and Wiss, Janney, Elster & Associates of Northbrook, Illinois. Meanwhile, the *Detroit News* retained a civil engineering professor from the University of Detroit to make a separate investigation.

The professor, Filomeno N. Almeida, declared in sensational front-page stories that MDOT had underestimated the damage and that the entire project was suspect. He cited evidence of water in conduits carrying post-tensioning tendons. He found exterior cracks. He said there were voids in the joints between the segments. He also declared that workmanship, which had been adequate at the start of the project, had deteriorated in the area where the accident occurred. MDOT engineering officials disputed each claim.

A year later, MDOT imported yet another expert, Peter Gergely of Cornell University's structural engineering department. Gergely recommended increasing the anchorage capabilities of the 28 pier footings carrying the greatest loads. This was done by enlarging them and installing reinforcing steel to connect the existing footings with new concrete surrounding them. This procedure was estimated to increase repair costs by $400,000.

In summer 1983 the state settled with its contractors. The state agreed to pay $4.8 million in claims and purchase several million dollars in specialized equipment from the contractors for future use, while the contractors agreed to pay $29 million in claims.

GAO blames original designers

The next year, the "official" cause of the accident was finally published in a report from the General Accounting Office (GAO), which blamed the original designers for "failure to recognize or adequately consider the effective load bearing capacity of the expansion joint as designed." The report noted that the designers approved a total load of 63,700 kips per foot (a kip is 1,000 pounds) but had failed to account for a 30-ton work platform at the end of span. The GAO stated that "handling manual procedures were approved which did not specify [the] requirement that the platform be removed."

In another part of the report, the GAO maintained that the contractor misplaced one leg of the launching girder by 4 feet, adding 2,600 kips per foot to the load. The GAO report only raised more disputes. H. Hubert Janssen of Indianapolis, consultant to ZCE (as the BVN/STS design firm was now known) insisted that two legs, not one, had been out of place and that the additional load was 5,400 kips per foot. He said that the work platform should not have been there at the time and that the state, which was supervising the job, should have known this. However, ZCE went out of existence when its contract with MDOT expired in February 1984 and was supplanted by HNTB.

Impact

The repair of the damaged section presented several challenges. Before repairs to the footing could begin, the base of the pier had to be stabilized to prevent vibration, which would cause a catastrophic collapse. The only way to repair the bearings at the top of the columns was to lift the superstructure first. MDOT estimated that the repairs would cost between $1 million and $3 million.

An elaborate monitoring system was installed that would warn of imminent collapse by detecting motion in the structure, and a system of cables was stretched between the broken superstructure and a counterweight weighing 1,100 tons placed below the broken expansion joint to prevent further shifting.

Unstable soil requires freezing

The first priority was to repair the crippled pier, which was supporting some 5,500 tons and was bearing on the fractured footing. Because unstable soil conditions could cause the pier to shift while the new footing was being installed, a complex job of freezing the soil beneath and around the pier began. Paul W. Steinberger, engineering manager of Geocentric, the freezing subcontractor, described the job as "the most complex freezing job that's ever been done in this country." It involved drilling an intricate network of 200 refrigeration pipes beneath the footing.

Raising superstructure allows footing to resume vertical position

Once the soil was stabilized, holes were drilled through the frozen earth 90 feet to bedrock. These holes were used as forms for three new foundation columns on each side of the footing. Steel reinforcing cages and concrete filled the holes to ground level. Then the outer parts of a new footing were placed around the new columns. The inner part of the footing would be installed once the superstructure was raised from the pier, allowing it to slip back into its vertical position.

A structural steel frame bearing on the new footing was then erected. Fitted with twelve 600-ton hydraulic jacks at the top, the frame was used to lift and carry most of the superstructure's load while the new footing, 11 feet thick, was cast around the base of the columns. After the concrete cured, the damaged bearings at the top of the column were replaced by new ones coated with Teflon and stainless steel to permit rotation and sliding. Then the superstructure was lowered to rest on the pier, and a set of hydraulic rams at the expansion joint slowly drew the superstructure back into alignment.

Northbound lanes open December 1987

General construction of the Zilwaukee Bridge resumed in March 1984, with S. J. Groves & Sons as contractor. Instruments were installed to monitor any unwarranted movement, and all parties had to agree whether each construction procedure was appropriate before any part of the work could proceed. Five years behind schedule and nearly 75 percent over budget with a total cost of $132.6 million, the Zilwaukee Bridge was finally opened to northbound traffic in December 1987 and southbound traffic nine months later.

Where to Learn More

Arnold, C. J. "Salvaging the Zilwaukee." *Civil Engineering* (April 1986): 46–49.

"Box Girder Wrenched out of Shape." *Engineering News-Record* (September 9, 1982): 10–11.

"Great Zilwaukee Bridge Mistake." *Construction Contracting* (April 1983): 14–15.

"More Problems Cited for Zilwaukee Bridge." *Engineering News-Record* (February 3, 1983): 13.

"Zilwaukee Woes Are Still Mounting." *Engineering News-Record* (August 2, 1984): 13.

Schoharie Creek Bridge collapses

Near Fort Hunter, New York
April 5, 1987

Two long-standing bridge engineering problems—scour and lack of redundancy—converge to destroy a New York State Thruway bridge. Ten people lose their lives.

Background

New York City and Buffalo—at opposite ends of the state of New York—are linked by the 550-mile-long New York Thruway. For 31 years the toll road operated without structural problems in any of its 819 bridges—until disaster struck April 5, 1987. The center span of the 540-foot Schoharie Creek Bridge collapsed without warning at 10:45 A.M., killing ten people as their vehicles plunged into Schoharie Creek.

Creek 20 feet deeper than usual

The highway, formally named the Governor Thomas E. Dewey Thruway, crosses the Schoharie Creek about 40 miles west of Albany, near the town of Fort Hunter. At 30 to 45 feet wide, the "creek" could more aptly be called a river. That rainy April day, flooding from the storm in the rural area seemed manageable, even though the Schoharie was running 30 feet deep under the 80-foot-high bridge, a good 20 feet more than usual. Thruway officials remembered that the Schoharie and other bridges in the area had survived a devastating flood in 1955, and they discounted the impact of some flood control measures being taken nearly 40 miles upstream.

Operators of the Blenheim Gilboa Pumped Storage Power Project feared damage to the reservoir, so they released flood waters into the Schoharie. The flow was later reported at a record 65,187 cfs (cubic feet per second). The previous record had been 55,000 cfs during the 1955 flood, which inundated Fort Hunter under 5 to 6 feet of water. This time the town stayed dry, because a new levee had been constructed along its north side.

Nearby residents reported hearing a loud explosion as the center spans of the Schoharie Creek Bridge collapsed without warning, killing ten people.

The levee kept the creek within its banks, but this intensified the flow. Witnesses later said that the water "seemed to boil over on the rocks below and tear at tree trunks leaning out over the banks of the creek."

As bridge collapses, at least five vehicles plunge into creek

The high flow caused state highway officials to close a bridge about a half mile downstream from the Thruway, but other bridges in the region were left open to traffic. On one of them, also just downstream from the Thruway, a 15-year-old boy became the main eyewitness to the collapse at 10:45 A.M. "All of a sudden I heard this noise. Water splashed up high, and the road and the concrete just fell in," he told reporters. He saw a truck plunge straight into the creek, as well as two cars go in, a small car that went in upside down and a large white car that was later found downstream with its top ripped off. At least two other cars also disappeared. Nearby residents said that they heard "an explosion," and ten minutes later "it sounded like a bomb going off" when the second span fell. The two spans were swept some 80 feet downstream, and the others dropped a few hours later when their piers failed.

State trooper Peter Persico had crossed the bridge a short time before but turned back after witnesses called the police. He found several cars and trucks parked on the partly collapsed bridge, "their drivers gawking," and with other troopers, he evacuated the bridge just before the second section collapsed. By nightfall police had located the two lost cars and a tractor trailer, but the rescue efforts were curtailed by high water and the threat of further collapse.

Bridges designed for 100-year storms

Thruway officials were astounded. Chief engineer Daniel S. Garvey declared: "We were just totally shocked. We're nonplussed about the whole thing. We don't understand it."

James A. Martin, deputy executive director, claimed that the Thruway's bridges were designed to withstand 100-year storms. He ordered an immediate inspection of all 819 Thruway bridges, and 100 were inspected the first day. By the end of that day three separate investigations had been arranged, and Senator Daniel Patrick Moynihan had flown in by helicopter, promising to ask the federal Department of Transportation (DOT) for $10 million to rebuild the bridge. (The Schoharie Creek Bridge had only been insured for its construction cost of $8.8 million, with a $2.5 million deductible.)

At the time of its collapse, the bridge was carrying an average of 16,000 vehicles per day. The 5-span bridge was built in 1952. The reinforced concrete deck sections each rested on a pair of steel girders about 7 feet high, and 100 feet to 120 feet long. The transverse floor beams were 55 feet wide.

Below were 4 piers consisting of 2 columns supported by a single shallow spread footing. The columns, 9 feet wide at the bottom and tapering to 7 feet wide at the top, were connected near their tops by a horizontal beam, giving the pier an H shape. Two piers were planted in the creek bed, 2 on the banks, and the superstructure also rested on 2 abutments.

Engineers vow to monitor water flow at bridge supports

The bridge had been rehabbed five years earlier and had been inspected last in April 1986, according to Robert C. Donnarums, Thruway deputy chief engineer. In the yearly inspections before that, no problems with the piers had been found, though in 1977 one inspector had recommended replacement of the riprap, which seemed to be missing on the upstream sides of the bridge. Riprap consists of large broken stones that are piled around the pier bases to protect them. Rehab plans made between 1981 and 1982 called for repairs and repaving of the four-lane deck but included no mention of the riprap. Donnarums added that now, after the collapse, investigators "would pay close attention" to the flow of water around the bridge supports.

Although they had records of rehab plans, Thruway and state officials soon learned that other records were in disarray. Construction details were lacking, and they had no idea how many other Thruway bridges were of the same steel plate, two-girder design. When they located the records, they found that no underwater inspection had been made, even though it had been urged at the time of the rehab.

Details of the Collapse

Investigators initially blamed the failure on the bridge's lack of redundancy (lack of backup load-bearing systems in case of component failure). Each of the 5 spans was supported independently, and the 2-girder design provided no alternative load paths in case one horizontal member failed. The lack of frame-action reinforcement in the concrete piers was also considered as a factor contributing to the suddenness of the collapse. But within two or three days of the collapse, investigators began to focus almost exclusively on hydraulics—the behavior of flowing water—and its effect on pier foundations. Dr. David Axelrod, state health commissioner and head of the State Disaster Preparedness Commission, held that there was "a whole series of unusual factors" about flows in Schoharie Creek.

Pier tipped into hole created by "scour"

Investigators discovered a report written by a state engineer reviewing the bridge design before its construction. The report warned that the proposed design did not take into account the frequent occurrence of ice jams in Schoharie Creek and the possibility of extremely severe floods. The engineer, advising that the 540-foot-length of the bridge would be too constrictive of the river, recommended lengthening the span to 775 feet.

The engineer's report was disregarded, and bridge construction and levee erection narrowed the river channel enough to greatly increase the stream's velocity. The rapid flow of water, in turn, had caused a great deal of erosion.

Scour washed out 9 feet of streambed

After the collapse, inspectors discovered that 9 feet of the streambed had been washed out from under the upstream side of one of the piers. This phenomenon is called scour, which is the erosion of the streambed by the force of the currents. When the pier finally tipped into this hole, the spans of the bridge above also disengaged from their supports and fell into the river. According to some experts, had the bridge been constructed with concern for redundancy—with 1 continuous span across the river instead of 5 independent spans—the girders would have merely sagged, not collapsed, at the location of the pier failure. The slump of the bridge would have provided a visual warning that the bridge was in danger of collapsing. There would have been enough time to close it and prevent any loss of life.

Lack of redundancy and reinforcements may have contributed to the severity of the collapse, but scour was blamed as the primary cause. Scour was not unknown before the Schoharie Creek Bridge collapsed, and it was not unique to that particular bridge. During that same spring, scour caused 17 other bridges to fail in the northeastern United States. The Hatchie River Bridge on U.S. Route 51 in Tennessee failed two years later, when riverbed material was lost from around a pier because the channel shifted laterally. Scour is also not easy to track. One reason is that the same water that scours out the streambed promptly refills the hole with loose silt and gravel. Without probing, even a diver cannot see what has happened, and in time, scour recurs.

Bridge officials ignored suggested maintenance

Two forensic engineering firms were brought in to determine the cause of failure: Wiss, Janney, Elstner & Associates of Chicago, and

Mueser Rutledge of New York City. The National Transportation Safety Board (NTSB) also made a full investigation. Their reports led the New York State Commission of Investigation to fix blame on the New York Thruway Authority (NYTA) by early August, only four months after the collapse. The commission based its judgment on the NYTA's failure to follow up on its engineering consultant's "critical" recommendation—to reinstall riprap around the piers—and the failure of a staff inspector to look at the footings during the annual inspection. They also pointed out that NYTA did not carry out its insurance company's advice to conduct an underwater inspection.

The commission concluded, "While the bridge design and construction may well have been deficient, with proper inspection and maintenance, this bridge would not have collapsed." The commission recommended a complete overhaul of the state's bridge inspection program, urged bridge inspectors to review drawings before site visits—a practice not rigorously enforced—and urged the state legislature to intervene by funding bridge repairs for needy localities.

The final verdict came in April 1988, just a year after the collapse: the National Transportation Safety Board blamed the NYTA for not maintaining the riprap. The report also cited "ambiguous plans and specifications for the original construction, an inadequate inspection and inadequate oversight by NYDOT and FHWA," the Federal Highway Administration, an agency of the U.S. Department of Transportation.

One witness saw a truck and two cars plunge straight into the creek. At least two other cars also disappeared.

Impact

Reconstruction of the bridge began soon after the collapse. Traffic was jamming all other available roads through the area, and Thruway officials hastened to retain the New York City firm of Hardesty & Hanover, which redesigned the bridge for quick reconstruction as well as for durability. The 14 plate girders are continuous over the piers and are spaced a conservative 8 feet apart. Designing the girders to be very stiff allowed the contractor to pour the concrete deck span by span rather than wait for all the steel on the entire bridge to be in place.

Bridge reconstructed to be unaffected by scour

A speedy solution to area-wide traffic congestion meant that new pier foundations had to be designed and constructed before the cause of the collapse was officially decided. The foundations were specifically designed to "make the system independent of scour." Steel-plated boxes, called sheetpile cofferdams, extend from the surface to 23 feet below the river bottom—8 feet below the foundation bottoms—to permit the construction crew to pour the pier foundations in dry conditions. The replacement bridge has deeper foundations and steel H-piles, which extend down to bedrock and rigidly connect to the foundations. With proper riprap, the design protects the pier bases from scour.

The Schoharie tragedy prompted new concerns and new regulations nationally. In a 1988 issue of *Civil Engineering*, Charles H. Thornton of Thornton-Tomasetti, a New York City engineering firm, published a "powerful plea for better bridge inspection practices." After studying the Schoharie disaster, he said that a bridge inspection program should require that the inspector have a specified degree of experience and familiarity with bridge design. He also recommended periodic evaluation of each bridge's complete structural and foundational system and numerous other measures to ensure that this type of failure did not occur again.

A 1989 study disclosed that 494 of the 823 U.S. bridge failures from 1951 to 1988 occurred as a result of hydraulic conditions, primarily scour of foundation material. Most of these collapses, however, were on local road systems and did not generate national news as the Schoharie Creek Bridge failure did.

Nationwide search for scour is launched

The search for scour developed rapidly. In 1988 the FHWA required states to identify bridges most likely to be vulnerable to scour damage and to set priorities for correcting the damage. Each state was to develop its own plan. The FHWA provided procedures for predicting potential scour depths and for designing countermeasures, but the agency underscored a need for more research and especially development of measuring devices. These are just now coming out of various laboratories.

The state of New York was among the first to bring in an independent task force to devise its plan. The task force evaluated nearly 20,000 bridges, then devised a simple, straightforward method for scour screening and priority procedure. Interim countermeasures installed at several sites included riprap and guide banks to withstand scour action and underpinning supports. Bridges were also ordered to be closed during storms.

New York's Department of Transportation, which once—according to one critic—had to "rummage through boxes in the basement" to locate specific bridge records, created the position of state bridge inspector general to bring order to the archives and the inspection process. The inspector general is also charged with reviewing new bridge designs.

Where to Learn More

"Can You Top This? (Schoharie Creek Bridge Reconstruction)." *Civil Engineering* (January 1988): 61–62.

Green, Peter. "Feds Blame State Agency for Schoharie Failure." *ENR* (May 5, 1988): 16.

Huber, Frank. "Update: Bridge Scour." *Civil Engineering* (September 1991).

Lane, Kate. "Schoharie Bridge Probe Yields Inspection Changes." *ENR* (May 14, 1987): 11.

Murillo, Juan A. "Scourge of Scour." *Civil Engineering* (July 1987): 66–69.

"Schoharie Report Offers Guidance on Inspection." *ENR* (December 10, 1987): 11.

Thornton, Charles H., and others. "Lessons from Schoharie Creek." *Civil Engineering* (May 1988): 46.

"Thruway Blamed for Collapse." *ENR* (August 13, 1987): 13–14.

V

Buildings and Other Structures

MGM Grand Hotel fire

Las Vegas, Nevada
November 21, 1980

Background

Sparks from a short circuit started a fire in the MGM Grand Hotel in Las Vegas, Nevada, on November 21, 1980. Despite having passed inspections, the building's fire protection systems could not prevent the fire from engulfing the world's largest gambling hall in smoke and flames. Thick black smoke filled the air ducts and the escape stairwells in the 21-story guest tower. Scenes reminiscent of Hollywood disaster films became reality as guests were awakened by screams and smoke. Some raced to the roof or ran down smoke-filled stairwells. Others called for help or huddled by broken windows as smoke billowed into their rooms. Eighty-five people died and over 600 were injured, primarily due to smoke inhalation rather than the fire itself. The disaster dramatically accelerated the updating of fire code regulations for both new and existing high-rise buildings.

Eighty-five people perish in the second-deadliest hotel fire in U.S. history, leading to nationwide reevaluation of fire codes.

Fire concentrates in casino and guest tower

An enormous facility, the MGM Grand Hotel features a casino, two 1,000-seat theaters, 40 shops, 5 restaurants, 2,076 guest rooms, a jai alai fronton, and a sports arena. A T-shaped tower of guest rooms rises above two entertainment levels, each as large as 20 football fields. The fire was concentrated near the casino on the upper entertainment level and in the guest tower. Of the fire protection systems designed and built into the MGM Grand, the most important during the fire were the building egress system, the fire suppression and alarm systems, the system of fire zones, and the heating, ventilation, and air conditioning (HVAC) system.

Helicopters rescued nearly 1,000 people from the fast-moving MGM Grand Hotel fire.

The hotel's building egress system was intended to provide safe routes of passage out of the building to safety. People in the casino and adjacent spaces could exit the building directly onto the street through several large sets of doors. Hotel guests could exit through the two stairwells in each of the three branches of the T-shaped guest tower.

No sprinklers in deli, where fire started

The fire alarm and suppression systems were designed to detect and control a fire. The alarm system would alert the hotel's security office of an outbreak of fire. The security office would then verify the alarm and use the building's alarms and public address systems to alert occupants to any danger. Fire suppression was provided by the automatic water sprinklers located in selected portions of the building; the heat of a fire would activate these sprinklers. The minimum guidelines of the fire code in effect at the time of the hotel's design did not require that the entire facility be equipped with sprinklers. Thus, the casino, deli, most floors in the guest tower, and many other areas were not protected by sprinklers. An especially important area that was not protected was the casino's "eye in the sky," a network of walkways above the casino's ceiling that allowed security personnel to monitor gambling operations unnoticed.

A system of fire zones isolated distinct sections of the hotel and casino facility to prevent the spread of fire and smoke and to protect occupants and property. Fire zones are formed by enclosing parts of a building with fire-resistant construction. These enclosures must effectively seal off the many openings between building parts and any gaps between construction materials through which fire and smoke can pass. Important openings in fire zone enclosures at the hotel include access panels into elevator and air shafts, structural joints for building movement, and the junctures of structural steel and drywall.

The heating, ventilation, and air conditioning (HVAC) system at the hotel was designed to stop the flow of air into the building during a fire. Fire dampers—hinged louvers similar to those on attic exhaust fans in many residences—are fundamental to this strategy. The fire dampers in large HVAC systems are typically installed adjacent to air distribution fans and where air crosses fire zone enclosures. Fire dampers are usually held open by a link designed to melt in the heat of a fire, thereby closing the louvers and stopping the flow of air. Thus fire dampers allow the passage of air through the building under normal circumstances and yet maintain the integrity of fire zone enclosures.

Modifications and materials defeat hotel's fire safety systems

Three factors caused the fire protection systems at the MGM Grand Hotel to perform poorly during the November 21 fire:

- The original construction did not fulfill the intent of the designers

- The contents of the casino level were unusually flammable

- Physical and operational modifications that were made since the opening of the hotel reduced the effectiveness of the fire protection systems.

Many construction deficiencies impaired the building's ability to suppress a fire. Potential smoke paths across fire zone enclosures were left open or were covered with materials that could not resist fire or stop smoke. Moreover, materials with inadequate flame-spread and smoke ratings were used to build ceilings and attic areas, and the elevator shafts were not adequately vented. The decorations, furnishings, and finishes on the casino level contained high amounts of synthetic materials. These materials can easily catch fire and burn, and they produce large amounts of smoke. Some fire dampers had been bolted open or had their "melt-down" links replaced with metal wire, which prevented the dampers from functioning as intended. In addition, several areas designed for 24-hour use—spaces where the fire code did not require sprinklers, on the assumption that any fire would be detected quickly—were closed down during low-use periods. One such area was the deli, where the fire started.

Despite safety flaws, MGM Grand passed fire inspections

The MGM Grand Hotel was thus a facility whose faulty and inadequate fire protection systems had been crippled by modifications. Despite the numerous flaws in its fire protection systems, the hotel was not in gross violation of the fire design and construction standards that prevailed at the time. The building underwent all required fire reviews and inspections before opening, and it passed a fire inspection just six months before the disastrous fire.

Details of the Fire

About 7:10 A.M. on November 21, 1980, an MGM Grand Hotel employee noticed sparks coming out of a gaming board in the closed-down deli. A chef tried to extinguish the fire but soon was forced to flee the area. The fire fed on fresh air supplied by the HVAC system and the abundant combustible materials, generating smoke that was laden with unburned fuels. These flammable gases collected in the "eye in the sky" above the casino and were infiltrating the casino space itself at the time that the fire department entered. Fire fighters noticed a stratified layer of smoke 6 to 8 feet from the ceiling. Twelve seconds later the smoke dropped to 4 feet above the floor, when the glass doors at the west end of the hotel exploded. The fire then became a fire storm.

The decorations, furnishings, and finishes on the casino level contained high amounts of synthetic materials—materials that caught fire easily and produced large amounts of smoke.

Only one system worked: automatic sprinklers

The inadequacies in the hotel's fire protection systems aided the progress of the fire. The fire incapacitated the alarm system, preventing those in the tower from receiving any warning. The fire zone enclosures failed, allowing smoke to enter the hotel's air ducts, elevator shafts, and stairways and to migrate throughout the building. The HVAC system, rather than depriving the fire of oxygen, provided it with fresh air. The chimney effect created by the vertical passageways and the HVAC system drew the smoke quickly up into the guest tower hallways and rooms. The building egress system, however, proved to be the deadliest factor. The only path of exit from the guest rooms was down the stairwells to the street. Once hotel guests entered these stairwells, self-locking doors prevented them from returning to the hallways. Thus, as smoke filled the stairwells, many guests became trapped. Only the automatic sprinkler system performed as designed, preventing the fire from spreading into the guest tower or beyond the casino-level areas that were not equipped with sprinklers.

Most of the guests and employees on the casino level were able to flee the building ahead of the fire. Some guests in the hotel tower, rather than attempting to evacuate down the stairwells, escaped to the roof. Many other guests remained in their rooms and broke or opened windows to gain access to fresh air. The fire department controlled the fire by 8:30 A.M. and evacuated most survivors by 11:00 A.M.

Impact

The most important impact of the fire in the MGM Grand Hotel was the nationwide revision of local fire codes. The fire brought two issues to the forefront of fire code discussions:

- The danger of smoke over and above that of fire, and
- The failure of fire protection systems to evolve with fire protection capabilities.

Emphasis shifts from saving property to saving lives

Given that over 90 percent of those who died in the disaster were overcome by smoke rather than fire, several communities increased the emphasis on smoke control in their fire protection regulations. These changes represented a shift in priorities from saving buildings to saving lives. HVAC system modifications included smoke detectors to signal the HVAC system. It would then exhaust air out of a fire area and pressurize the surrounding spaces to contain the circulation of the smoke. Fire protection professionals disagree over the effectiveness of the various smoke control methods, but many cities now require that all high-rise buildings be equipped with smoke control systems.

Fire-protection systems in buildings lag behind fire protection evolution for two reasons: technologies continuously improve, and buildings have long lives. In the case of the MGM Grand Hotel, for example, eight years lapsed between design and construction of the hotel and the November 21 fire. In that period, new advances in fire protection for high-rise buildings were incorporated into local fire codes, including smoke detectors, refuge centers, direct connection of fire alarms to fire departments, communication systems for use by firefighters during a fire, and smoke exhaust systems. The long life span of buildings causes many facilities to remain in use long after their fire protection systems have become outdated.

Retrofitting regulations are unpopular with building owners

Fire codes can require existing buildings to be upgraded to comply with newly adopted guidelines, but such retroactive codes, while increasing safety, carry large costs for building owners and communities. The higher cost of "retrofitting" and operating high-rise buildings can ultimately reduce the attractiveness of a city to developers and businesses. In 1980, few fire codes were retroactive. Since the MGM Grand fire, several cities have adopted retroactive codes. The legality of several of these fire code provisions has been challenged in the courts.

However, the MGM Grand Hotel fire illustrates that technological disaster can catalyze social change. Disasters often shift the relative values placed on different aspects of the community—human life, buildings and facilities, economic expansion. The nation's second deadliest hotel fire provoked fire code reforms and hints at the political significance of the MGM Grand Hotel disaster.

Where to Learn More

Brannigan, Vincent M. "Record of Appellate Courts on Retrospective Fire Safety Codes." *Fire Journal* (November 1981): 62–72.

"Cities Stiffen Fire Codes." *Engineering News-Record* (January 14, 1982): 16–17.

"Compromised Codes Threaten Life Safety." *Engineering News-Record* (January 29, 1981): 26–30.

"Fire at the MGM Grand." *Fire Journal* (January 1982): 19–32.

"MGM Fire Rekindles Smoke Debate." *Engineering News-Record* (December 4, 1980): 14–15.

"MGM Fire: Short Circuit to Disaster." *Engineering News-Record* (June 11, 1981): 13–14.

"MGM Grand Hotel Victim of Old Code." *Engineering News-Record* (November 27, 1980): 10–11.

Hyatt Regency Hotel walkways collapse

Kansas City, Missouri
July 17, 1981

Two suspended walkways collapse and kill 114 people. Caused by a "simple" design change, this disaster is the worst structural failure to date in the United States.

Background

On the evening of July 17, 1981, some 1,500 people gathered in the glass-walled, four-story atrium at the Kansas City Hyatt Regency for the hotel's weekly tea dance contest. As the Steve Miller Orchestra played swing music, some contestants danced on the floor of the open lobby while others watched from three pedestrian walkways that crossed the space overhead. Just as the band struck up Duke Ellington's "Satin Doll," a sound like a clap of thunder rang out, and two of the overhead walkways collapsed, showering glass, concrete, steel, and people down on the dancers below. It took rescue personnel using heavy equipment 12 hours to reach the last victims trapped under debris. The collapse, which has since become known as one of the greatest structural tragedies in the history of the United States, killed 114 people and injured nearly 200 others.

Design of hanger rod did not come from the approved drawings

The collapse of the Hyatt walkways was caused by a failed hanger rod–box beam connection. The connection that failed was not built according to original architectural drawings. Instead the connection was constructed according to an alternative design prepared by the steel fabricator. Investigators blamed the failure of the walkways on that design change, which effectively doubled the stress on the hanger rod–box beam connection. Engineers later remarked that the design problem was so basic that it could have served as an exercise in an undergraduate engineering class.

Firefighters (upper right) perform rescue efforts as others look on. The tragic collapse of the walkways killed 114 people and injured nearly 200 others.

The tragedy focused attention on:

- Engineering details being designed by fabricators and materials suppliers
- Responsibility of the project's engineer-of-record for the oversight of such details

Two engineers eventually lost their professional licenses as a result of their role in the walkway collapse.

Dramatic atriums are trademark of Hyatt hotels

The Kansas City Hyatt Regency was designed by a joint venture between architect Patty Berkebile Nelson Duncan Monroe Lefebvre Architects Planners and the structural engineering firm of Gillum Colaco Associates. It opened about one year before the walkways collapsed.

The hotel included a 40-story guest room tower and a 4-story wing containing restaurants and meeting rooms. The wing and tower were connected by a large, open atrium lobby. Dramatic atriums had become a trademark of Hyatt hotels during the preceding ten years, following a trend begun with the opening of architect John Portman's Atlanta Hyatt Regency, which contained a spectacular 22-story-high open atrium.

Details of walkway construction

The Kansas City Hyatt atrium contained three hanging walkways, each 120 feet long—consisting of four spans of 30 feet each—and slightly more than 8½ feet wide. The walkways spanned the atrium at the second-, third-, and fourth-floor levels. Each walkway was connected to balconies at either end. The walkways' box beams were welded to embedded plates to form a rigid joint on the south end, while the north end connection was constructed as an expansion joint. Sliding bearings provided the beams with room to expand or contract—in reaction to changes in temperature—without causing stress or cracking.

The walkways were constructed of lightweight concrete on metal decking. Each walkway was supported from underneath along its length by 16-inch-deep, wide-flange steel beams, and across its width by 8-inch-deep steel box beams, welded beneath the walkways at 30-foot intervals. The hanging rods that suspended the walkways from the roof supports attached to both ends of the box beams at these seams. The box beams were made from pairs of 8-inch steel channels that were welded together at their open ends. The hanging rods, also made of steel, measured 1¼ inches in diameter.

The fourth- and second-floor walkways were constructed one above the other along the west wall of the atrium, while the third-floor walkway was built about 13 feet away toward the center of the room. To support the weight of the walkways, the ends of the hanging rods passed through holes that were drilled through the welded box beams and secured underneath by a washer and nut. No stiffeners, plates, or other reinforcements were used to further secure the connections. Each walkway was hung from three pairs of steel rods.

Approved design called for steel rods to be 45 feet long

The design change that occurred after the original drawings were approved concerned the way in which the walkways were connected to the steel hanging rods. The original drawings called for both the fourth- and second-floor level walkways to be hung from the same continuous

The third walkway of the Hyatt Regency remained intact, while two other suspended walkways collapsed—primarily due to a change in the design of the suspension rods.

steel rods. The rods, about 45 feet long, would have been attached at the top to the atrium ceiling, passed through the box beams at the fourth-floor level walkway, then continued through the box beams at the second-floor level walkway. The weight of each walkway would have been supported by the nut-and-washer connection under the box beams at each level.

Instead of using continuous rods, however, contractors hung the walkways from two separate sets of rods. In the modified construction, each transverse box beam of the upper walkway had two holes drilled through either end of it: one hole 2 inches from the end and another hole 6 inches from the end.

- One set of rods passed through the outside holes, which suspended the fourth-floor bridge from the atrium ceiling and supported its weight with only a nut and washer underneath each beam.

- Another set of rods passed through the inner holes. These rods were held in place by a nut and washer connection above the fourth-floor beams, and they supported the second-floor walkway from the bottom by another nut and washer connection.

The weight of the second-floor walkway was originally designed to be supported by the ceiling connections. The design that was actually implemented diverted the weight of the second-story walkway from the ceiling and channeled it instead to the rods and beams of the fourth-floor walkway.

Henry Petroski, a civil engineering professor at Duke University who studied the failure, suggested that the situation could be compared to one in which two men hang separately by their hands from the same rope. If their hands are strong enough to support their own weight, the men are not in danger of falling. But if instead of grasping the rope, one man holds onto the legs of the other man, the man at the top would be supporting twice his weight. If he is just barely strong enough to support the double load, any small weight addition would be enough to make him lose his grip.

The design was probably changed because the use of such a long steel rod in the detail would have been somewhat tricky. Perhaps it would have required casting a long rod in which the lower half was threaded like a screw, so that a bolt could be twisted almost halfway up its length to support the upper walkway. Engineers later suggested that the best way to accomplish the original design would have been to use a sleeve nut below the fourth-level walkway, a device that could connect two separate threaded rods to form one continuous rod.

Details of the Collapse

Initial speculation about the tragedy contended that such a large crowd of people had congregated on the walkways to watch the dance contest that the weight capacity was exceeded. Others theorized that people dancing on the walkway had set in motion a vibration that caused the collapse.

National Bureau of Standards blames disaster on design change

Investigations into the collapse were launched by several groups, including Failure Analysis Associates of Palo Alto, California; Packer Engineering Associates of Chicago; Lee Lowery, a professor at Texas A & M University; Rex Paulsen of Fay Engineering Company in Denver; and—probably the most prominent—the National Bureau of Standards (NBS). The NBS simulated the conditions of the collapse in the laboratory and tested the actual materials from the collapsed walkways as they became available.

The report produced by NBS investigators, led by Edward O. Pfrang, the chief of NBS structures division, came out in February 1982. The NBS blamed that one design change for the entire catastrophe. As built, NBS investigators claimed, the walkways could support only 27 percent of the combined dead load and live load of the Kansas City building code.

- "Dead load" is the weight of the materials of the bridge.
- "Live load" is the weight of the people standing on the bridge.

TV crew was taping when the walkways collapsed

A local television crew had been filming the tea dance, which provided investigators with video footage to aid in their analysis. Investigators estimated that 63 people were watching the dance from the second- and fourth-floor walkways at the time of the collapse. This number would have created a live load of less than 10 percent of the weight the walkway should have been able to support. Each of the hanging rods should have been able to support 68,000 pounds to meet the design load requirements. Instead the box beam connections collapsed under a load of just 18,600 pounds.

The National Bureau of Standards (now National Institute of Standards & Technology) determined how the walkways collapsed:

1. The welding of one of the fourth-floor box beams split apart.

2. Then the nut and washer supporting the bottom of the fourth-floor walkway slipped through the hole in the box beam.

3. The combined loads of the fourth- and second-floor walkways then transferred to the remainder of the fourth-floor connections.

4. The remaining fourth-floor box beams also fractured, unable to sustain the weight.

5. The fourth-floor walkway then collapsed onto the one below it, sending both walkways crashing down onto the lobby floor.

The NBS report conceded that the connections as originally designed would have withstood the loads that caused the modified design to fail. Nevertheless, the original design was lacking as well, according to the NBS investigators, who calculated that the original hanging rod design would have supported only about 60 percent of the design load requirement under the building code. The NBS also had some minor reservations about the quality of the welding of the channels that constituted the box beams, but the agency concluded that materials and workmanship did not play a significant role in the disaster.

Impact

Within days after the disaster, Crown Center Redevelopment Corporation, which owned the hotel, removed the remaining atrium walkway at the third-floor level. Investigating engineers and lawyers representing victims of the catastrophe objected to the walkway's removal.

Inspectors find signs that third walkway is also damaged

However, the owners were vindicated for removing the walkways when a later inspection of the third-floor walkway revealed that these box beams were deformed near the hanger rod–box beam connections. Apparently the third-floor walkway was also overstressed. The side walls of these box beams were bowed out, and the bottom walls were deformed by as much as ¼ inch where the washer and nut pulled up against the bottom of the beam.

The hotel's owners replaced that walkway with a new 17-foot wide causeway at the second-floor level, supported on ten columns built up from below. Closed for 11 weeks during the repairs, the hotel reopened in

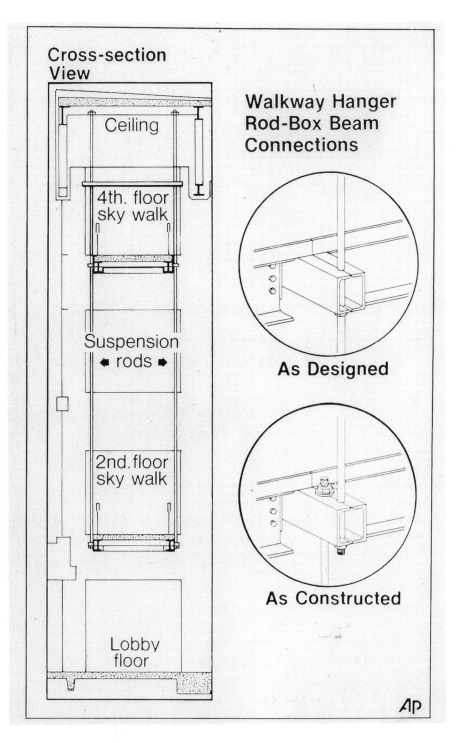

Cross-section View

Ceiling

4th. floor sky walk

Suspension ← rods →

2nd. floor sky walk

Lobby floor

Walkway Hanger Rod-Box Beam Connections

As Designed

As Constructed

AP

This cross-section of the Hyatt walkways shows the faulty design that was a primary factor in the tragic collapse.

fall 1981. After the Kansas City Hyatt disaster, the owners of the nearly 60 other Hyatt hotels with atriums of various designs ordered structural safety inspections for their facilities.

At least 150 lawsuits were filed in circuit court seeking total compensatory damages of $1 billion and punitive damages of about $500 million. Virtually all of the suits were eventually settled out of court. Citing a lack of evidence, a 1983 grand jury issued no indictments for criminal negligence against the parties involved.

Missouri architectural board brings formal complaints against main project engineers

In February 1984, however, the Missouri Board for Architects, Professional Engineers, and Land Surveyors filed complaints of gross negligence in the design and analysis of the walkways against the building's chief structural engineer, Jack K. Gillum, and the project engineer, Daniel M. Duncan. Both Gillum and Duncan were at that time officers with St. Louis–based GCE International. The subsequent hearing was supported by 450 exhibits and produced 5,000 pages of transcript and was closely watched by the entire engineering community.

Gillum and Duncan argued that the failed connection was designed by the project's steel erector and fabricator, Havens Steel Company of Kansas City. Gillum and Duncan's attorney asserted that it was common practice to delegate the task of designing steel connections to the steel fabricator, because the fabricator could best select the most economical way of building such details. The engineer for Havens Steel explained that the detailing on the Hyatt project had been subcontracted to WRW Engineering of Kansas City. Neither Gillum and Duncan nor Havens Steel admitted to changing the connection design, and each blamed the other for the alteration. However, the shop drawings submitted by Havens Steel were stamped and initialed by the architectural joint venture (Patty Berkebile Nelson Duncan Monroe Lefebvre Architects Planners and structural engineering firm Gillum Colaco Associates) and the contractor.

Much testimony at the hearing concerned a collapse of part of the atrium's roof during the construction of the hotel in 1979, long before the walkways collapsed. Although no one was hurt in the roof collapse, it prompted the building's owners to ask for a review of the structure, essentially giving the engineers a second chance to catch the walkway design mistake. Evidence showed that Duncan then submitted a report stating that he had checked the suspended bridges. Gillum stated that he told the project's construction manager that every connection in the atrium had been reviewed.

Engineers lose engineering licenses permanently

In November 1985 Gillum and Duncan were found guilty of negligence for having failed to review the steel shop drawings. Missouri administrative law judge James B. Deutsch, who presided at the hearing, censured the engineers for deliberate fraud in not thoroughly checking the connections after the 1979 roof collapse. In January 1986 the Missouri Board for Architects, Professional Engineers, and Land Surveyors voted unanimously to permanently revoke the engineering licenses of Gillum and Duncan and to revoke the certificate of authority for their firm, GCE International.

The Hyatt tragedy attracted widespread public attention because so many people perished. The disaster also forced the engineering community to reconsider the meaning of the professional seal that an engineer affixes to a set of project documents. Judge Deutsch insisted that while the courts would allow the engineer to delegate the job of designing a detail to another party, the responsibility for that detail remained with the engineer who affixes the professional seal to the documents.

More than a decade later, the lines of responsibility between the structural engineer-of-record and the project fabricator are still a subject of controversy. Although the Hyatt case seemed to indicate plainly that the structural engineer could be held responsible for errors made by a fabricator, many structural engineers and architects do not want to be held liable for errors made by others and have tried to "write away" responsibility in project specifications and other contractual documents. Many firms also require the fabricator's own engineer to seal shop drawings that come out of that company.

Although some long-standing issues remain unresolved, the disaster had impact on the engineering profession:

- Numerous seminars, conferences, and papers kept open the debate in professional circles.

- The American Consulting Engineers Council instituted a peer review process.

- The American Society of Civil Engineers brought out its *Quality in the Constructed Project* manual in 1988.

Where to Learn More

Alm, Rick. "Hyatt Engineers Lose Licenses in Missouri." *Engineering News-Record* (January 30, 1986): 11.

"Death Trap in Kansas City." *Newsweek* (July 27, 1981): 30–31.

"Hyatt Hearing Traces Design Chain." *Engineering News-Record* (July 26, 1984): 12–13.

"Hyatt Hotel Engineers Cited for 'Negligence.'" *Engineering News-Record* (February 9, 1984): 14.

Marshall, R. D., and others. *Investigation of the Kansas City Hyatt Regency Walkways Collapse.* U.S. Department of Commerce, National Bureau of Standards, 1982.

Petroski, Henry. *To Engineer Is Human.* New York: St. Martin's Press, 1985.

Pfrang, Edward O., and Richard Marshall. "Collapse of the Kansas City Hyatt Regency Walkways." *Civil Engineering* (July 1982): 65–68.

Ross, Steven S. *Construction Disasters.* New York: McGraw-Hill, 1984.

Ocean Ranger rig sinks

Atlantic Ocean, 160 miles off the coast of Newfoundland, Canada
February 15, 1982

Background

The *Ocean Ranger*, a mobile offshore drilling rig in the North Atlantic, combined the fateful ingredients for disaster: control operators not properly trained to handle emergency situations, lifeboats unsuited to the weather conditions, substandard emergency protective clothing, freezing water temperatures, and a fierce storm. With all those ingredients in place, a small broken window in the rig's control room eventually caused the entire structure to capsize on February 15, 1982. None of the crew members was saved.

Seawater short-circuited the control panel

No one knows how the window broke. The heavy winds, high waves, or possibly the day's drilling operations could have been responsible. Regardless of how it happened, the missing window allowed seawater to splash in, drenching some of the control panels. Then the ballast control panel short-circuited, which caused the rig to lean precariously forward.

Owned and operated by Ocean Drilling and Exploration Company (ODECO) of New Orleans and Mobil Oil Canada, the *Ocean Ranger* was a semisubmersible oil-drilling rig built in 1976 in Hiroshima, Japan. The working platform, supported on eight columns, resembled an eight-legged table. The two groups of four legs were atop multicompartment tanks called ballast tanks. The 400-foot-long ballast tanks would be filled with air to keep the structure floating or with water to help the rig maintain its stability. If heavy loads were placed on the front of the *Ocean Ranger*, it would list in the front. The listing would be countered by allow-

A broken window on the offshore oil-drilling rig in the North Atlantic precipitates a chain of events that capsizes the rig. All 84 crew members perish.

Eighty-four crew members perished as the *Ocean Ranger*'s ballast control panel short-circuited, causing the rig to capsize.

ing water to flow into the rear ballast tanks, which would level the rig. After the front load was removed, the rear ballast tanks would be pumped out to restabilize the rig. Since proper ballast system operation was crucial to oil rig stability, it was mandatory that the control room operator be experienced.

The ballast control panel was customized for the *Ocean Ranger*. The switches for opening and closing the valves on the ballast tanks operated electrically, but if power were lost, a manual control rod could perform the same function. However, without electricity the gauges would not work and the position of the valves could not be determined.

Details of the Disaster

The weather on February 14, 1982, was relatively calm during the day. The *Ocean Ranger* and two other mobile offshore drilling rigs were drilling for oil 160 miles off the coast of Newfoundland, Canada. Ranging from 300 feet to 400 feet in length and operating in waters at least 250 feet deep, these rigs had an inherent potential to become unstable. Because of this ever-present danger and other possible hazards, each drilling rig was required to be accompanied by a rescue ship. However, the rescue ship typically doubled as a supply boat and drifted within two miles of the drilling rig.

The following sequence of events was reconstructed from radio transmissions received by the other drilling rigs and rescue ships in the vicinity, because the tragedy that claimed the *Ocean Ranger* left no survivors.

Short-circuited controls are "opening and closing on their own"

The ill-starred chain of events began at dinnertime on February 14. Internal communication was overheard that the crew was mopping up water and cleaning up broken glass in the control room and also that the crew members were getting shocks from a control panel. During the evening the weather worsened considerably. The drilling rigs were pummeled by 55-foot-high waves, 80-mile-per-hour winds, and subzero temperatures, yet conversations throughout the evening between the *Ocean Ranger* and the other ships showed no cause for alarm. Around 9:00 P.M. an internal transmission on the *Ocean Ranger* requested that an electrician report to the control room, because valves for stabilizing the drilling rig were opening and closing on their own.

Shortly after midnight the *Ocean Ranger* radio operator sent out a distress message. The message, which was repeated 10 to 20 times in a half hour, transmitted that the drilling rig had a severe list and required immediate assistance. Less than two hours later the crew of the *Ocean Ranger* proceeded to life raft stations to begin abandoning ship, because the listing of the oil rig had reached the point of no return. As the crew boarded the lifeboats, they most likely believed that in a short time they would reach safety. Unfortunately, the lifeboats offered these crew members little chance of survival.

Lifeboats offer no security in rough seas

The lifeboats the crew boarded in the early morning were not designed for the fierce environment they were about to battle. Launching the boats required them to be dropped 70 feet into the ocean. This drop and the high winds made safe use of these emergency boats a fantasy. Some investigators later concluded that some of the lifeboats in all likelihood sustained extensive damage from bouncing off the side of the *Ocean Ranger* as they were launched.

Probably only one of the lifeboats successfully survived the launch and had a chance for rescue. The bow of this lifeboat was damaged during the launch and water poured in through the hole. Yet the crew seemed to have a good chance of being saved: they wore life jackets, their lifeboat was still afloat and its motor was operating, and rescue boats were approaching, but the rescue ships lacked adequate lifesaving equipment.

Rescue ship is eight miles away

During the severe weather, the rescue ship *Seaforth Highlander* had drifted eight miles away from the *Ocean Ranger*. Rescue ships typically stay within a mile or two of their assigned ship. This distance is close enough for emergency response, but still far enough away to avoid collision and tangling of anchor lines. When the *Seaforth Highlander* heard the distress signal of the *Ocean Ranger*, she proceeded full speed ahead—but did not arrive until an hour later.

The crews on the rescue ships for the other two mobile offshore oil-drilling rigs had been monitoring the *Ocean Ranger* situation. At 1:15 A.M. they were desperately needed. Within one to two hours both rescue ships were in the vicinity of the sinking oil rig; but the *Seaforth Highlander* was the only rescue team that saw any living members of the *Ocean Ranger* crew.

Just before 2:00 A.M., the *Highlander* sighted distress flares. As the crew approached the flares, they spotted a lifeboat with a hole in it heading in their direction. The lifeboat and the rescue ship maneuvered to be side by side. The rescue ship tossed lifelines to the lifeboat. The lifeboat was fastened to the rescue ship with one of these lines. While the rough sea violently tossed the two boats about, several men exited the covered part of the lifeboat to board the rescue ship. Suddenly the lifeboat capsized and threw the men into the sea.

Ship has no rescue basket to scoop up frozen men

Within minutes the *Seaforth Highlander* had thrown inflatable life rafts to the men in the sea. They landed right next to the men, but they made no effort to grab for them. The bitter cold seawater had exerted its paralyzing effects on the crew. There was little else the *Seaforth Highlander* could do, because the ship was not equipped with baskets to scoop people out of the water. After many attempts to pull the crew out of the ocean, the *Seaforth Highlander* had to move on in search of other crew members, but in vain.

The search for survivors continued by air and sea for a week after the rig capsized. Drift plots were calculated to estimate the farthest distance that an object could drift from its original position under existing weather conditions. The search within the drift plot boundaries located several lifeboats and bodies, some more than 40 miles away from the oil rig. The search was called off on March 1, 1982, with a death toll of 84 people. Only 22 of the bodies were ever recovered.

The lifeboats the crew boarded in the early morning were not designed for the fierce environment they were about to battle. Launching the boats required them to be dropped 70 feet into the ocean. This drop and the high winds made safe use of these emergency boats a fantasy.

Impact

The window that broke in the control room was at the side of the ballast control panel. The incoming water short-circuited the panel, which meant the system could only be operated by manual controls. In all likelihood not a single person on the *Ocean Ranger* knew how to operate the ballast control panel manually.

Even if an operator was experienced with the manual controls, the system was not realistically operable since all information gauges were also shorted out. A radio transmission from the *Ocean Ranger* stated that the ballast valves were opening and closing on their own. The erratic behavior of the control board was causing the ballast tanks in the front of the rig to fill with water, giving the *Ocean Ranger* a list of 10 to 15 degrees forward in rough seas. This angle of tilt would normally not be an immediate problem. However, due to the rough seas, the openings to the anchor chains started to flood, drastically increasing the list. The *Ocean Ranger* was now totally out of balance and starting to sink into the North Atlantic, so the crew abandoned ship.

Drilling rig was not equipped with protective clothing

Investigators had the distinct impression that damage could have been limited to loss of property only. They felt that all crew members probably reached the lifeboats safely, but the boats themselves were not designed to withstand the pounding they would get in a 70-foot drop from the rig into the ocean. Also, many crew members floating in lifejackets could easily be spotted, because the jackets were equipped with emergency lights. But without protective clothing from the cold ocean, the jackets were merely keeping dead bodies afloat. And those who managed to keep their lifeboat afloat did not fare any better. The rescue boats were not equipped to pull survivors from them under the harsh storm conditions.

As a result of the capsizing of the *Ocean Ranger,* the National Transportation Safety Board instructed the U.S. Coast Guard to require the following conditions:

1. An operating manual must be provided for semisubmersible mobile offshore units that includes guidance for countering accidental flooding of ballast tanks as well as for preventing flooding into the anchor chain compartments.

2. Control room operators must be licensed in ballasting procedures for mobile offshore drilling units.

3. Exposure suits to protect crew in the ocean must be provided for 150 percent of the persons on offshore drilling rigs in locations where water temperatures may go below 60°F.

4. Vessels assigned for rescue purposes to the mobile offshore drilling rigs must have capabilities for retrieving disabled persons from the water under adverse conditions.

5. The suitability of approved lifeboats, life rafts, and lifesaving equipment must be evaluated.

One month after the disaster, the families of the victims brought a $1.7 billion lawsuit against ODECO and Mobil Oil Canada, the owners and operators of the *Ocean Ranger*. All but a handful of the families had reached out-of-court settlements with the companies involved.

Where to Learn More

"Capsized Rig's Design O.K." *Engineering News-Record* (February 25, 1982): 12.

Clugston, Michael. "An Aftermath of Sorrow and Anger." *Maclean's* (March 1, 1982): 28–31.

———. "A Preventable Tragedy." *Maclean's* (August 27, 1984): 45.

———. "How the *Ocean Ranger* Sank." *Maclean's* (February 21, 1983): 14.

———. "The Search of the *Ocean Ranger.*" *Maclean's* (May 3, 1982): 17–18.

Giniger, Henry. "Hope Fades for 84 on Rig; 18 on Soviet Freighter Die." *New York Times* (February 17, 1982): 3.

Joyce, Randolph, "The Cruel Sea." *Maclean's* (March 1, 1982): 26–27.

———. "The *Ocean Ranger*'s Mystery Deepens." *Maclean's* (March 8, 1982): 24–25.

LeMoyne, James, and others. "The *Ocean Ranger*'s Night of Death." *Newsweek* (March 1, 1982): 48.

National Transportation Safety Board. "Capsizing and Sinking of the U.S. Mobile Offshore Drilling Unit *Ocean Ranger,* February 15, 1982." *Report No. NTSB-March 83-2.* Washington, DC: Government Printing Office, March 1983.

"Sea Extracts Its Price in Hunt for Oil." *U.S. News and World Report* (March 1, 1982): 6.

"Wreck of the *Ocean Ranger.*" *Time* (March 1, 1982): 29.

VI

Nuclear Plants

SL-1 reactor explodes

Near Arco, Idaho
January 3, 1961

Background

Three technicians die in the nation's first nuclear reactor accident to cause loss of human life.

Near Arco, Idaho, on the afternoon of January 3, 1961, three technicians died in an explosion at the Experimental Stationary Low-Power Nuclear Reactor, an experimental reactor known as SL-1. Mechanical, material, documentation, and operator failures all played roles in the nation's first reactor accident to result in human lives being lost. Lethal levels of radiation contaminated the site for months afterward, and the accident dampened the American public's overly high expectations about nuclear power.

Postwar America: From nuclear weapons to power generation

During World War II the U.S. Atomic Energy Commission (AEC) concerned itself exclusively with the production of nuclear weapons. After the war the AEC turned its attention to peaceful applications of nuclear energy—such as production of electric power. One of the locations it selected for experimentation was the area around Arco, Idaho.

Idaho wasteland is ideal site for nuclear experimentation

Arco, then a tiny town of 780, attracted the AEC's attention for the 400,000 acres of wasteland that surrounded it. Void of plants, animals, and people, the area seemed an ideal place to conduct nuclear experiments. AEC officials reasoned that if an accident were to occur, there would be little danger of damaging either environment or human life. The largest city of any size, Idaho Falls was a "safe" 65 miles to the east.

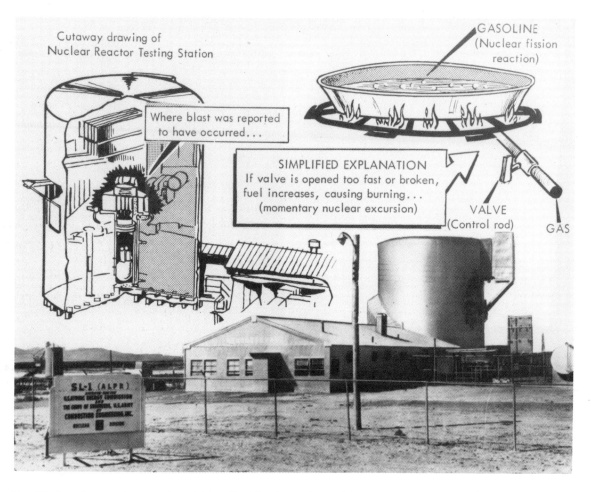

Cutaway drawing of
Nuclear Reactor Testing Station

Where blast was reported
to have occurred...

GASOLINE
(Nuclear fission
reaction)

SIMPLIFIED EXPLANATION
If valve is opened too fast or broken,
fuel increases, causing burning...
(momentary nuclear excursion)

VALVE
(Control rod)

GAS

SL-1 (ALPR)

The SL-1 nuclear reactor near Arco, Idaho. Contributing to the accident were vague explanations in the operating manual and poorly designed control rods, which allowed operators little room for error when inserting and removing them from the reactor core.

Much of the area around Arco contained porous lava beds, a feature AEC officials regarded as another plus. Nuclear engineers believed these beds would be useful as a place to dump radioactive waste resulting from their experiments.

The first experimental reactor, called Experimental Breeder Reactor No. 1, or EBR-1, began operating at the Arco site by the end of 1951. The Arco site later became known as National Reactor Testing Station (NRTS). On December 20, 1951, when the reactor was first turned on, it provided enough electricity to light four 200-watt bulbs. The reactor was taken to full power the next day. It began supplying electricity to the reactor building itself and the equipment associated with it.

First reactor melts down

The success of EBR-1 was an exciting step forward for Atomic Energy Commission engineers. But the plant was in operation only four years before experiencing a meltdown of its core. The reactor was shut down in 1955 and was never returned to service.

SL-1 goes online in 1958

Meanwhile other construction was going on at NRTS as engineers studied a number of possible reactor designs. One such experimental reactor was the Stationary Low-Power Reactor No. 1, or SL-1. The reactor was designed and built by scientists from the Argonne National Laboratory outside Chicago. Those scientists had earned a reputation as the most reliable and dependable workers in the field. SL-1 was first put into operation in August 1958. Responsibility for its construction and operation was shared by four different entities: the Atomic Energy Commission, the U.S. Army, Phillips Petroleum Company, and Combustion Engineering.

The purpose of SL-1 was to test the feasibility of a small nuclear reactor for supplying electricity in remote areas such as the Arctic. Under full operation the plant eventually produced 3 megawatts of power and seemed to be a completely successful experiment.

Details of the Accident

Firefighters at the NRTS Fire Station and Security Headquarters, eight miles from SL-1, first learned of a problem when an automatic alarm sounded on January 3, 1961. Six AEC (Atomic Energy Commission) firefighters, an assistant fire chief, and a security officer immediately drove to the reactor. Their drive took nine minutes.

Dosimeters indicate radiation at lethal levels

The arriving safety workers saw no sign of fire or accident. They methodically worked their way from building to building at the reactor complex as they checked for smoke or fire. When they approached the reactor itself, the needles on their personal dosimeters jumped off the scale. The

REM is an acronym for Roentgen Equivalent Man. It denotes the dose of radiation that will cause the same biological effect as one roentgen of X ray exposure.

dosimeters they were wearing, or REM (Roentgen Equivalent Man) meters, were set to measure a radiation level of 25 roentgens. The rescue crew turned back at this point, because they had no instruments to measure the true level of radiation in the reactor. A second crew arrived with meters that could read 200 roentgens. Their instruments also read off scale.

Radiation exceeds 500 roentgens

Finally a third crew reached the site. The crew consisted of a firefighter and a health physicist from the Materials Testing Reactor operated by Phillips, about ten miles away. The meters carried by these workers read to a maximum of 500 roentgens, far beyond the level expected to occur in an accident. Again the needles went off the scale. The roentgen is used as a unit of exposure dose, and a single dose of 500 roentgens is fatal for half of all exposed individuals.

Rescue workers knew that radiation within the reactor was at lethal levels. Their search of the other buildings had failed to turn up the three maintenance men (two soldiers and a sailor) who were supposed to be working on the reactor. It seemed likely that they were inside the reactor building—but were they still alive?

At 10:35 P.M., four officials from Combustion Engineering and the Phillips health physicist began going into the reactor room to search for the three military workers. They stayed no more than three minutes at a time to limit their total exposure to radiation.

Missing workers found—one pinned to ceiling by reactor rod

The rescue team found one worker dead on the floor of the reactor building and a second barely alive. He was removed but died on his way to the hospital. The third man was eventually found pinned to the ceiling of the reactor room by one of the control rods. It had penetrated his body from groin to shoulder. His body was removed six days later through the efforts of a rescue team and a remote-controlled crane.

In all nuclear reactors, a liquid or gas flows through the core of the reactor and heats up. Its purpose is to take away the heat generated by fission in the reactor core, so it is called a coolant. The heat is transferred from the reactor to a steam generator, where it boils water to steam and functions from then on like any power station powered by coal. SL-1, like most nuclear reactors, used water as a coolant. The SL-1 reactor vessel that surrounded the reactor was 4.5 feet in diameter and 14.5 feet high. It resembled a silo. The reactor contained long, vertical fuel rods that held pellets of uranium. The reactor also contained control rods, similar in shape to the fuel rods. The control rods were spaced between the fuel rods to absorb neutrons and thus, by being raised or lowered, keep the fission reaction going steadily.

Operator's manual didn't give full explanation

Although SL-1 was a well-designed reactor in many respects, it appeared to have at least two fatal flaws. The first involved the core's control rods. When fully immersed in the core, the rods absorbed so many neutrons that they completely halted the chain reaction. Full withdrawal was so dangerous that it was never permitted, because when the control rods were fully withdrawn, the chain reaction would go out of control and produce core meltdown. The margin for error was a small one. The instruction manual for the reactor advised workers to remove the rods by "not more than four inches," but it did not explain why this was important.

On January 3, however, workers were faced with two problems. First, the reactor had been shut down for 11 days for routine maintenance. The control rods had been disconnected from the automatic chain drive to which they were usually attached. Workers had to manually pull the rods up far enough to reattach them to the drive.

Control rods unexpectedly warped

That task would not have been too difficult except for the second problem. The channels holding the control rods and the rods themselves had warped slightly. The rods no longer slid smoothly in and out of the housings. Authorities believe that a worker must have had to use unusual force to pull up on one of the control rods. These experts think that when the rod finally broke loose and began to move, it slid more than four inches out of the core. For a fraction of a second, excess neutrons bombarded the fuel rods. Then the core temperature spiked up a few thousand degrees, heated the cooling water into steam, and blew off the reactor lid. The whole accident probably took less than a second to occur.

Reactor overloaded with uranium

A second design element may have contributed to the disaster. SL-1 contained enough uranium to keep the reactor running without refueling for three years. The fuel supply overtaxed the ability of the control rods to regulate the reaction. The fix for this problem was to attach thin strips of boron to the aluminum surface of the fuel rods. Boron soaks up neutrons

The Atomic Energy Commission firefighters who first reached the SL-1 reactor wore dosimeters that measured up to 25 roentgens. Since their dosimeter needles were past the maximum capability of 25 roentgens, the firefighters knew the radiation in the area of the SL-1 reactor exceeded 25 roentgens. Radiation exposure accumulates in the human body and is known to cause cancer.

much as the cadmium does in the control rods. Thus the boron strips provided the reactor with additional capacity to control neutrons in the core.

Proper operation depended on jerry-built boron strips

What no one realized, however, was how easily the boron strips tear loose from the fuel rods. When scientists were eventually able to study the bottom of the reactor, they found pieces, strips, and flakes of boron that had come loose from the fuel rods. For some time (no one knows exactly how long) the extra neutron-absorbing capacity provided by the boron strips—as well as the extra safety margin—was diminishing. The "four inch caution" suggested in the operator's manual was based on the presence of the boron. Far from being an actual fix, attaching the boron strips created a false sense of security.

Impact

Many design and operation mistakes led up to the January 3, 1961, explosion at SL-1. On the plus side, however, the decision to locate the reactor test site in a remote area was wise. The fatality possibilities were enormous, but the only fatalities to result were the three military workers at the plant at the time of the explosion. There were no other apparent injuries, and no long-term health effects were ever reported by the workers who entered the reactor in the first hours after the accident. Fourteen of these workers received radiation doses of more than 5 roentgens; six, more than 20 roentgens. Radiation levels were monitored downwind from the accident, and there were measurements of iodine-131 detected in the air, in vegetation, and in milk. However, this contamination was reported not to have exceeded the maximum permissible levels for continuous exposure to the off-site population.

American public was led to expect too much

The major impact of the accident was its effect on people's attitudes about the commercial applications of nuclear power. Since the late 1940s scientists and government officials had been painting an overly optimistic picture of the role of nuclear power in the world of the future. The public was told that it could look forward to atomic-powered ships and airplanes and a virtually limitless supply of safe, clean electricity from nuclear power plants. And for at least a decade the nuclear power industry appeared to be keeping that promise. Although relatively minor problems were reported from time to time, more and more commercial plants were coming on line.

Once burned, public is twice shy

The SL-1 disaster, the nation's first reactor accident to cause human lives to be lost, cast a pall onto people's high hopes and the glowing promises of nuclear energy. Official investigation of the accident indicated that neither Washington nor the local Atomic Energy Commission office was aware of design errors in the reactor. The AEC was also unaware of the control rod problems that were accumulating.

After the January 3 accident the Atomic Energy Commission closed down the only other reactor, located in Greenland, designed like SL-1. It also reaffirmed its decision to locate nuclear power plants only in non-urban areas. AEC scientists eventually developed a power-to-distance ratio that quantified and watered down this policy. They figured that a 1.5 megawatt plant would have to be built at least 17.7 miles from a large urban area (recall that in the early 1950s Idaho Falls, 65 miles from the 3-megawatt SL-1 reactor, was judged a "safe" distance).

Blame was heaped liberally on one of the reactor's co-owners, Combustion Engineering. Investigators pointed out that the company already knew enough about the reactor's problems to have closed it down long before the accident. Combustion Engineering did not disagree with this conclusion.

Design flaws fixed easily

The design problems that caused the SL-1 accident were actually relatively simple to correct:

1. Officials discontinued the practice of overloading the core with uranium. The downside of this change meant they could not operate the reactor as long before refueling. The upside of the change meant they would not have to use the problematic boron strips.

2. Officials forbade removal of the center control rod. Investigators found that keeping a rod in the center of the core at all times was critical. Removing just that one rod, regardless of the position of the other rods, could in and of itself induce a nuclear accident. Had the three military workers dealt with a warped

The SL-1 disaster cast a pall over people's high hopes and the glowing promises of nuclear energy. Official investigation of the accident indicated that neither Washington nor the local AEC was aware of design errors in the reactor. The AEC was also unaware of the control rod problems that were accumulating.

rod in any of the other positions, the core meltdown might never have happened.

3. Finally, officials ordered the installation of automatic stops on all SL-1 reactors. These stops would physically prevent removing any of the control rods more than their safe maximum distance.

Where to Learn More

Armagnac, Alden P. "The Atomic Accident that Couldn't Happen." *Popular Science* (September 1961): 52.

"Hot Stuff." *Newsweek* (February 27, 1961): 64–65.

Lapp, Ralph E. "A Small Atomic Accident." *Harper's* (June 1961): 47–51.

Norton, Boyd. "Atomic Roulette." *Chemtech* (July 1980): 412–415.

Three Mile Island reactor melts down

Middletown, Pennsylvania
March 28, 1979

Background

On March 28, 1979, a partial meltdown occurred at Three Mile Island nuclear power plant in Middletown, Pennsylvania. A valve in a pipe that carried cooling water to the reactor was supposed to be in the closed position. When it accidentally stuck in the "open" position, it allowed cooling water to continue flowing out of the reactor. Heat built up within the reactor core. This caused some fuel rods to melt and released radioactive gas and water.

Operator mistakes add to disaster

A number of operator mistakes contributed to the accident. While no one was killed or injured, expert opinion remains divided about possible future health problems that may result from the partial meltdown. The Three Mile Island accident became a critical factor in the dramatic retreat from nuclear power that has occurred in the United States since 1979.

The Three Mile Island nuclear power plant was built in two parts, Unit 1 (TMI-1) and Unit 2 (TMI-2). Both units were constructed on a small island in the Susquehanna River in south-central Pennsylvania, near Harrisburg. Unit 1 was put into operation in September 1974. Construction on Unit 2 began in 1968 and was completed ten years later. Unit 2 was formally declared open December 30, 1978.

Three electric companies own TMI

Both units at Three Mile Island were built by Babcock & Wilson, a

The reactor core of the nuclear power plant suffers partial meltdown. This is the worst accident in the history of American commercial nuclear power generation.

The Three Mile Island complex, containing two reactors and four cooling towers. Unit 2 was the site of a major nuclear accident in 1979.

giant construction company that specializes in nuclear power plant components. The nuclear complex is owned by three power companies: the first two, Pennsylvania Electric Company and Jersey Central Power and Light Company, each hold 25 percent of the stock; Metropolitan Edison holds the remaining 50 percent.

Pressurized-water reactors provide electricity to 346,000

Both units at Three Mile Island are pressurized-water reactors (PWR). When neutrons bombard uranium atoms in the fuel rods at the core of such reactors, fission results. Fission is the splitting of atoms, a reaction that produces great amounts of energy in the form of heat.

The Unit 2 plant at Three Mile Island (TMI-2) had 36,816 fuel rods, each filled with hundreds of pellets of enriched uranium metal about the size and shape of small checkers.

Fission heat eventually drives turbines

The heat produced by the fission reaction is removed by cooling water that is pumped through the reactor core. The cooling water is then pumped out of the core to a heat exchanger, where it is used to boil water in a secondary system. The steam produced in the secondary system drives turbines, which turn electric generators.

At their peak capacity, the Three Mile Island plants produced 880 MW (megawatts) of electricity. This amount was sufficient to meet the needs of 346,000 residents of surrounding Berks, Lebanon, and York counties. Pressurized-water reactors of this design are among the most common of all reactors in use in the United States today.

Constructed with fail-safe design elements

Like all nuclear power plants, the Three Mile Island reactors were constructed to protect even if an accident were to occur:

- The core itself is enclosed in a steel casing almost 9 inches thick.

- The core and cooling system, in turn, are enclosed within a large containment dome, 190 feet high and 140 feet in diameter. Walls of the dome are 4 feet thick and made of reinforced concrete.

- The containment dome exists to capture any gases, radiation, or other materials released during a leak or accident in the core or cooling system.

Problems plagued TMI-2 operations from beginning

As carefully designed as TMI-2 was, it experienced a number of problems almost from the day its license was granted in February 1978.

During the test period—

- Valves opened and closed without an operator initiating any order

- Valves remained stuck in the wrong position

- Seals broke

- Recording instruments failed to work properly

- Other "glitches" developed.

The top portion of the TMI-1 reactor, which is still in operation. The similarly designed TMI-2 reactor was the site of the 1979 accident.

The plant seemed jinxed

The situation did not improve even after the plant officially went on-line in December 1978. Only two weeks into operation, two safety valves failed. This caused the plant to shut down for two more weeks. During February 1979 additional problems with valves, seals, pumps, and instruments developed. The plant seemed jinxed by one mishap after another.

Problems with nuclear power generation were not unknown. More than a year earlier, a nuclear safety expert for the Tennessee Valley Authority, Carl Michelson, discovered safety problems with other Babcock & Wilson reactors. Michelson reported his findings to the Nuclear Regulatory Commission (NRC). The NRC monitors the safety of nuclear power plants in the United States. The NRC did not act on Michelson's report and did not request any modifications in Babcock & Wilson reactors.

Details of the Failure

The accident at TMI-2 began at 4:00 A.M. and lasted for more than five hours. During that time more than 40 distinct events occurred, some caused by mechanical error and some caused by human error.

The series of events began with a routine maintenance operation: the changing of water in a system of pipes. During the operation, air accidentally got into a pipe, cutting off the flow of cooling water to the reactor core.

Backup system was down for maintenance

Under normal circumstances, instruments that monitor this process would have detected this problem and automatically switched over to a backup cooling system.

However, the pumps used in the backup system were also undergoing routine maintenance—and thus could not deliver water to the pipes. Normal procedures call for operators to attach tags to the pumps to indicate that they are undergoing maintenance. These tags hung down in such a way that they hid the warning lights. Even though the system correctly indicated that a problem existed, operators could not see the warning lights.

Fuel rods begin to melt, reach 5,000°F

Without cooling water removing the heat from the reactor core, the core rapidly began to overheat. Its temperature eventually reached more than 5,000°F, and fuel rods began to melt.

As they melted, the fuel rods—now like lava—burned through the steel casing of the core and reached the floor of the containment vessel itself. Had the molten fuel melted through that barrier, a total meltdown could have occurred. The resulting explosion would have breached the containment barrier and released huge levels of radioactive material into the air.

Other plant systems fail in turn

During the meltdown, other components of the plant continued to fail. Water in the cooling system started to boil. This made an emergency relief valve automatically open to allow the water out into the containment dome. But the valve then failed to close, and even more water was lost from the cooling system.

Operators err

All the while, operators were:

• Reading gauges that were not working properly, or

• Reading working gauges but taking the wrong actions.

Meltdown of fuel rods is actually expected under such circumstances. But the formation of a huge bubble of steam and hydrogen gas inside the containment dome was unexpected. The steam came from coolant water boiling out of the core. But where did the hydrogen come from?

Meltdown causes water to produce dangerous hydrogen gas

The accident at Three Mile Island taught scientists something they had not realized before: that the radiation and heat released in the meltdown causes cooling water not only to boil but also to decompose into its separate molecules: oxygen and hydrogen. Hydrogen is a special source of concern, because it is highly explosive.

Although no explosion occurred at Three Mile Island, the hydrogen could have exploded. The resulting release of radioactive materials would have affected an area hundreds of square miles around the plant.

Radioactive gas and water escape

Nevertheless, radioactive material did escape from the reactor in other ways. Some contaminated coolant water was automatically pumped out of the containment dome into a nearby holding building—this water was eventually dumped into the Susquehanna River. Radioactive gases from the core also escaped though vents in the containment dome. The level of radiation was so high in these gases that it damaged the instruments designed to monitor them.

When they first learned about the TMI-2 incident, scientists and politicians at first seemed uncertain and confused. Two hours after the backup pumps first failed, the plant's chief engineer was told that a "minor snafu" had occurred at the plant. An hour and a half later the situation had escalated into a "general emergency" for Unit 2. By 8:15 A.M. plant managers set up a direct telephone line to Nuclear Regulatory Commission offices in Washington, D.C.

Officials had to revise press releases

In their early press conferences, plant officials said that less than 1 percent of the core's fuel rods were damaged and that the internal temperature of the reactor was only 2,000°F. Later they revised those figures to at least 50 percent loss of fuel rods and a temperature of 5,000°F.

Everyone involved seemed anxious not to cause unnecessary panic. At first Pennsylvania governor Richard Thornburgh wanted to order a general evacuation of the area. Later he decided more modest precautions were sufficient. He advised people to stay in their homes with their windows closed.

Radiation measured 30 millirems at dome

Radiation measurements taken the day of the accident showed levels ranging from 30 millirems immediately above the containment dome to 5 millirems three miles downwind from the plant. The problem is that experts still disagree about the health effects of low-level radiation. A person receives about 20 millirems of radiation from a typical chest X ray, so the levels measured at TMI-2 seem reasonably safe. But some authorities believe that each additional dose of radiation, no matter how small, increases a person's chance of contracting cancer later in life. Therefore it is possible that the TMI-2 accident may have some impact on long-term human health.

Cleanup efforts on the damaged reactor began almost immediately. The most important source of concern was the steam-hydrogen bubble in the containment dome. Engineers vented part of the gas into the atmosphere through filters that removed its radioactivity. They also were able to transfer some of the steam-hydrogen mixture to an outside building where the hydrogen was combined chemically with oxygen to make water.

Hot reactor core is cooled down

Cleanup crews removed the 20,000 gallons of radioactive water in the auxiliary building and transferred it to holding tanks. Finally, a variety of methods were used to cool down the hot reactor core itself and prevent any further meltdown.

The long-term cleanup of TMI-2 was a slow, dangerous, and complex process that took 11 years. Eventually 150 metric tons of damaged fuel rods and other reactor components were shipped to the Idaho National Engineering Laboratory for storage and analysis. The total cost of the cleanup process was $973 million.

TMI-2 will not be repaired

There are no plans to repair TMI-2. Safety experts will continue to observe and monitor the plant until early into the next century. At that point it will be decommissioned, as will its companion reactor (TMI-1). By then the total lifetime cost of nuclear power generation at Three Mile Island will be close to $2 billion.

Impact

The TMI-2 accident brought about changes at many levels. In the immediate area surrounding the Three Mile Island complex, residents expressed fears about the possible effects the accident might have on their health over the long range. Area residents have filed more than 2,200 lawsuits because of the accident. To date 280 claims have been settled at a total cost of about $14 million.

TMI-1 gets $95 million of modifications

Operations at Unit 1 were significantly changed by the events at TMI-2. More than 100 modifications were made to the older plant and cost $95 million. While the necessary renovations were being performed, the plant was shut down for six years. TMI-1 managers used the plant's downtime to develop and implement an improved and expanded training program for plant operators. Much of the training at the plant is now done in an $18 million full-scale replica of the TMI-1 control room.

Rogovin and Kemeny reports

Governmental bodies and industry groups also investigated the TMI-2 accident. Two studies are best known: the NRC (Nuclear Regulatory Commission) study, whose results became generally known as the Rogovin Report, and a study commissioned by President Jimmy Carter, eventually known as the Kemeny Report. In the wake of their investigations, the NRC issued new regulations in 1980 that required certain modifications to be made to existing plants and ordered new safety features to be included in future plants.

The nuclear power industry also concentrated more on safety issues in the aftermath of the TMI-2 accident. Companies established the Institute of Nuclear Power Operations (INPO) to review and evaluate nuclear power plant performance and to recommend ways of improving performance.

Antinuclear activists win a round

Probably the greatest single result of the TMI-2 accident was to provide a public forum about nuclear-generated power in the United States. Orders for nuclear power plants were already declining in the year preceding the TMI-2 event. But the accident seemed to mark a real turning point in the public's attitude about nuclear power. Activists around the nation used Three Mile Island as an "almost-worst-case" scenario. They spoke out against plants under development and worked to defeat plans for new plants.

The efforts of the antinuclear activists have been largely successful. Since 1979 no new nuclear plants have been ordered, and 59 planned reactors have been canceled. Seventeen new plants have been opened, but all were in final stages of planning or construction at the time of the accident.

Where to Learn More

Adams, John. A. "TMI plus 5; Part I: A Slow Comeback." *IEEE Spectrum* (April 1984): 27–33.

Burnham, David. "Nuclear Experts Reportedly Knew of Flaw in Some Reactors in 1977." *New York Times* (May 25, 1979): 18.

Eisenhunt, Darrell G. "TMI plus 5; Part II: NRC as Referee." *IEEE Spectrum* (April 1984): 33–39.

Fischetti, Mark A. "TMI plus 5; Part III: Band-Aids and Better." *IEEE Spectrum* (April 1984): 39–43.

Ford, Daniel. "Three Mile Island, 2: The Paper Trail." *New Yorker* (April 13, 1981): 46–47.

Franklin, Ben A. "Files Show Many Prior Problems at Three Mile Island." *New York Times* (April 19, 1979): 18.

Gray, Mike, and Ira Rosen. *The Warning: Accident at Three Mile Island.* New York: Norton, 1982.

Levine, Joe, Mary Hickey, and Denise Laffan. "Cleansing the Atom." *Time* (March 1989): 18–24.

Matthews, Tom, and others. "Nuclear Accident: Three Mile Island Plant, Pa." *Newsweek* (April 9, 1979): 24–30+.

Moss, Thomas H., and David L. Sills, eds. *The Three Mile Island Nuclear Accident: Lessons and Implications.* New York: New York Academy of Sciences, 1981.

Rogovin, Mitchell, and George T. Frampton, Jr. *Three Mile Island: A Report to the Commissioners and the Public.* 2 vols. Washington, DC: Nuclear Regulatory Commission Special Inquiry Group, January 1980.

U.S. President's Commission on the Accident at Three Mile Island. *Report of the President's Commission on the Accident at Three Mile Island: The Need for Change; The Legacy of Three Mile Island.* John G. Kemeny, chairman. Washington, DC: Government Printing Office, 1979.

Tsuruga radioactive waste spills

Tsuruga, Japan
March 8, 1981

Background

Tsuruga Nuclear Power Plant is in Fukui Prefecture in Japan. On March 8, 1981, a worker was cleaning out a pipe in the building where radioactive wastes were treated. An indicator light reported that the pipe valve was closed, but it was in fact open. As the worker flushed the pipe, the holding tank overflowed. Radioactive wastes covered the floor of the waste treatment plant, leaked into an adjacent building, seeped into the ground, and eventually worked their way into the Sea of Japan. Officials at the plant attempted to hide the accident from public notice, but evidence surfaced six weeks later during a routine study of seaweed in the area.

Due to human error, nearly 4,000 gallons of radioactive wastes escape from a nuclear power plant in Japan.

With few natural resources, Japan emphasizes nuclear energy

Japan has amazed the world by its economic development since World War II. The one weakness of the "Japanese miracle," however, has been that nation's lack of natural resources, particularly energy. The Japanese do not have sufficient reserves of coal, oil, or natural gas to maintain their pace of economic development. As recently as 1978 the nation imported 90 percent of its energy resources, 75 percent of which is petroleum.

The OPEC (Organization of Petroleum Exporting Countries) oil embargo of 1973 made it abundantly clear to the Japanese how dependent they were on imported energy sources. The government renewed its emphasis on developing nuclear power in order for Japan to be more energy independent.

Nuclear power in Japan, however, has a troubled history. With two of its cities devastated by the world's first two atomic bombs, Japan has been

The valve that leaked radioactive waste at the nuclear power plant in Tsuruga, Japan.

fundamentally concerned about nuclear energy, even for peaceful uses. When the Japanese government finally decided to permit and to sponsor nuclear research in 1955, it adopted a law containing many safeguards for citizens and guaranteeing that nuclear research will never be used for military purposes.

Japan's reactors supply 26 percent of nation's power

Commercial nuclear power development progressed rapidly after passage of the 1955 Basic Law on Atomic Energy. The first research reactor opened in 1957 at Tokaimura in Ibaraki Prefecture, and the first commercial plant came on-line eight years later.

The government originally planned to open more than 30 nuclear power plants by the year 1990, reducing the nation's dependence on foreign oil from 75 percent to 50 percent. That objective was overly optimistic, but progress was rapid nonetheless. By 1986 the Japanese opened 31 commercial reactors that produced almost 24 megawatts of electricity.

Atomic energy became the largest single source of power in the nation, accounting for 26 percent of the nation's energy supplies.

Tsuruga plant goes on-line in 1970

The nuclear power plant at Tsuruga, the nation's second oldest, opened in 1970. It was constructed by General Electric (GE) according to designs provided by Ebasco Services of New York State. In the aftermath of the waste spill at Tsuruga March 8, 1981, GE pointed out that the accident's location was in a part of the plant built by the Japanese, not by the American company. All Japanese reactors, including Tsuruga, are operated by the Japan Atomic Power Company (JAPC). JAPC's activities are monitored by two government agencies, the Agency of Natural Resources and Energy (ANRE) and the Ministry of International Trade and Technology.

Details of the Accident

A routine operation is the treatment of the nuclear waste generated by a power plant, which requires the periodic flushing out of pipes. On March 8, 1981, a worker at the Tsuruga plant was performing this operation. The indicator light reported that the intake pipe valve was closed, but it was in fact open, so that as the worker flushed the pipe, the radioactive wastes continued flowing into a holding tank until it began to overflow.

Valve indicator light malfunctions

A later investigation by ANRE (Agency of Natural Resources and Energy) found that the indicator light for the valve was not working properly. The light indicated that the valve was closed when actually it was still open. ANRE concluded, however, that the operator "should have known that [the] indicator light was malfunctioning."

The subsequent chain of March 8 events was not revealed to the public until April 18. Even plant managers themselves did not know about the spill until March 10. The story that gradually emerged changed with each new report.

Radioactive water overflows its tank

Apparently radioactive water overflowed its holding tank and covered the floor of the waste treatment building, which was connected to

The immediate result of the Tsuruga accident was to publicize the plant's past history of safety violations. In the months following the March 8 spill, JAPC officials gradually revealed details of more than 30 accidents in the plant's operating history.

the reactor building by pipes. Then it ran into the laundry room next door, where it drained through cracks in the laundry room floor and into the ground beneath. Eventually it diffused into a sewer line and worked its way into the nearby Urazoko Bay, an inlet of the Sea of Japan.

When plant officials learned of the accident, they ordered workers to clean up the radioactive wastes by hand, using mops and buckets. The job was an enormous one, taking more than two weeks and requiring the hiring of about 50 additional workers from outside the plant. According to later estimates, anywhere from 15 to 40 tons of radioactive wastes spilled out of the holding tank, making it Japan's worst spill. Authorities claimed that not more than one cubic meter of radioactive wastes ever reached the bay, an estimate regarded by others as dubious.

Plant officials keep spill a secret

Officials decided not to announce the leak to the outside world, but local health officials discovered evidence of it accidentally when they were making a routine check of seaweed in Urazoko Bay. They found levels of radiation at least 10 times greater than normal. The levels of some isotopes were even higher. The reading for cobalt 60, for example, was 5,000 times greater than any previously recorded in the area. Workers traced the source of radiation to an outlet pipe for the sewer running underneath the nuclear power plant.

Confronted on April 18 with this evidence, plant officials admitted to the accident of March 8. They claimed that the accident posed no danger either to workers in the plant or to nearby residents. Those involved in the cleanup, officials claimed, had been exposed to no more than 35 millirems of radiation, less than the level permitted under health regulations in both Japan and the United States. There is no way to verify their claim, however, because plant executives later conceded that workers' radiation-exposure badges were altered.

A REM—meaning Roentgen Equivalent Man—calibrates the dosage of ionizing radiation the human body has been exposed to. The average annual radiation dose received by a person in the United States is about 180 millirems. Most of this dose is due to natural radiation and medical and dental X rays.

Impact

The immediate result of the Tsuruga accident was to publicize the plant's past history of safety violations. In the months following the March 8 spill, JAPC officials gradually revealed details of more than 30 accidents in the plant's operating history. Prior to March 8 was an incident January 24–28, 1981. During repair of a cracked pipe, more than 40 workers may have been exposed to dangerous levels of radiation. On January 10, 1975, a spill similar to the March 8 accident released 13 tons of radioactive material into the surrounding environment and exposed 37 workers to radiation during the cleanup process.

Officials claim spill has no great significance

JAPC (Japan Atomic Power Company) officials continually tried to reassure the general public that the Tsuruga accident was minor and of no great significance. It was, they said, "nowhere near as serious as America's Three Mile Island," and they promised that there was never any "real damage" as a result of the spill. However, local residents were not easily reassured. "How do we know what is the effect of 15 tons in the bay?" an anxious citizen asked.

The plant was closed down for repairs, which cost about $10 million and kept the plant out of operation for six months.

The company's belated April 18 announcement drew criticism nationally. Japan's second largest political party, the Socialists, called for an end to the country's nuclear power program and a shutdown of all existing plants. A grass-roots antinuclear movement saw its membership increase by more than 45,000.

In response to the scandal, JAPC's chairman Tomiichiro Shirasawa and president Shunichi Suzuki resigned May 13. They accepted full responsibility for the accidents and cover-ups at Tsuruga and called for a return to confidence in the role of nuclear power in Japan's future.

Long-term adverse affects have been minor

The long-term effects of the Tsuruga accident have been relatively minor, especially compared with the disasters at Three Mile Island and Chernobyl. The role of nuclear power in Japan's energy equation has remained fairly constant. Japan's official policy is still one of weaning itself from dependence on foreign oil. As a result, the development of nuclear power has continued at about the pace the nation set for itself two decades ago.

Where to Learn More

"Earlier Mishap Is Revealed at Japanese Nuclear Plant." *New York Times* (April 21, 1981): 5.

Firth, Suzanne. "Chilling Reminder of Things Past." *Maclean's* (May 4, 1981): 32–33.

"Forty-five Workers Are Reported Exposed to Nuclear Radiation." *New York Times* (April 26, 1981): 6.

"Japan Says Nuclear Mishap Exposed 56 to Radiation." *New York Times* (April 22, 1981): 8.

"Japanese Concede Errors in Nuclear Plant Mishap." *New York Times* (April 4, 1981): A6.

"Japan's Three Mile Island." *Newsweek* (May 4, 1981): 42.

"Nuclear Contamination Found in Japanese Soil." *New York Times* (April 19, 1981): 7.

"Nuclear Energy Policy." *Kodansha Encyclopedia of Japan.* New York: Kodansha International, 1983.

"Nuclear Executives in Japan Resign over Recent Mishaps." *New York Times* (May 14, 1981): 11.

"Nuclear Power Plants." *Kodansha Encyclopedia of Japan Supplement.* New York: Kodansha International, 1986.

"Repercussions Continue at Japan's Three Mile Island and at the Original." *Power* (July 1981): 124–126.

Smith, R. Jeffrey. "Japanese Agitated by Nuclear Plant Spill." *Science* (June 5, 1981): 1124.

Stokes, Henry Scott. "For the Japanese, Sudden Misgivings about Nuclear Power." *New York Times* (May 16, 1981): 3.

Chernobyl reactor explodes

Chernobyl Nuclear Power Plant, Ukrainian SSR
April 26, 1986

On April 26, 1986, two mammoth explosions blew apart Unit 4 of the Chernobyl Nuclear Power Plant in the Ukrainian SSR (now Ukraine) of the former Soviet Union. The plant is located about 70 miles north of Kiev, the capital of the Ukraine. At least 31 workers and emergency personnel were killed immediately or died soon after the accident as a result of radiation sickness. Some 200,000 residents of the area were evacuated, and radioactive debris was carried by clouds over most of northern Europe. Total fallout from the accident eventually reached a level ten times that of the atomic bomb dropped on Hiroshima, Japan, during World War II.

The world's worst nuclear power plant disaster occurs when human errors lead to a dangerous heat and steam buildup, triggering two enormous explosions.

Background

The Chernobyl complex of four reactors was constructed between 1977 and 1983. By 1986 all four units were operating nearly at capacity and generating 4 million kilowatts of electricity. The reactors are graphite-moderated, water-cooled boiling water reactors known as RBMKs. At the time of the accident, the Soviet Union operated 21 reactors of this design, and together they generated about 15 percent of all the electricity in the nation.

RBMK reactors were chosen for weapons-making capabilities

The RBMK design is uncommon, and plants of this kind are found almost exclusively in countries of the former Soviet Union. They were built because they have the ability to perform two quite different func-

Unit 4 of the Chernobyl nuclear complex, which was the site of world's worst nuclear accident to date.

tions at the same time: generate electricity and produce plutonium. The decision to build these dual-purpose plants goes back to the 1950s. The Soviet government committed itself to a weapons-development program that would keep it on a par with the U.S. nuclear arsenal. Soviet weapons developers needed a constant and dependable supply of plutonium, which is the raw material needed for atomic and hydrogen bombs. The RBMK design was selected, therefore, because it could meet not only the domestic need for electricity but also the military demand for weapons-quality plutonium.

The core of the RBMK reactor consists of a huge pile of graphite blocks weighing about 2,000 tons. Imbedded in the block are about 1,700 fuel rods, each containing hundreds of disc-shaped pellets of uranium. The graphite slows down the production, or fission, of neutrons in the uranium. (Most other reactors use coolant water to perform this function.) The reactor also contains control rods, which dampen the nuclear reaction when they are lowered into the block.

The RBMK is cooled by water that is pumped through the graphite block, from bottom to top. The cooling water travels through channels in the block that also hold the fuel rods. As the water passes over the fuel rods, it picks up heat and begins to boil. Steam formed by the boiling water collects at the top of the reactor, where it is removed and delivered to the steam turbine to generate electricity.

Soviet reactors are a risky commercial choice

The RBMK reactor contains a number of design characteristics that make it risky to operate as a commercial power source. Western scientists had been warning Moscow about these dangers for a decade before the Chernobyl accident. Nevertheless, the Soviets continued to have confidence in the RBMK. This confidence was driven not only by the practical reality of military demands, but also by one important advantage of the RBMKs: they can be refueled without the extended downtime that Western reactors need to refuel. Ironically, this very advantage led to the disaster of April 26, 1986.

RBMKs have no containment shell

The RBMK design makes no provision for a containment shell, which all U. S. reactors use. A containment shell is a last-ditch safety feature built to retain the gases and radioactive materials that would be released during an accident in the reactor core. The Three Mile Island accident in 1979

The damaged reactor at Chernobyl is visible in the center of this photo, taken about two weeks after the 1986 accident. Eventually workers built a steel-and-concrete tomb to isolate the reactor, but by 1992 this shell cracked, and radioactive material again leaked into the environment.

would have been immeasurably more serious if the products of the core meltdown had escaped to the atmosphere. (See the entry "Three Mile Island reactor melts down" in this volume.)

Another feature of RBMK design is that a loss of cooling water increases the rate of fission (or nuclear splitting)—and hence heat production—in the core. The process is just the reverse of that in water-moderated reactors (such as those used in the United States), where loss of cooling water causes a decrease of power production in the core. This design flaw creates a peculiar situation: the RBMK reactor is most likely to go out of control when it is operating at lowest power.

Details of the Catastrophe

At 1:24 A.M. on Saturday, April 26, two enormous explosions rocked

Unit 4 of the Chernobyl Nuclear Power Plant. They blew the roof off the plant and spewed radioactive gases and debris more than 1,100 meters (3,600 feet) into the atmosphere. Two workers were killed instantly, and another dozen received such high levels of radiation that they died within the next two weeks.

Initial Soviet silence

As with other nuclear accidents on Soviet soil, government officials provided no public information about the nuclear accident. Only when monitoring instruments in Sweden detected a dramatic increase in wind-borne radiation did the Soviets finally acknowledge the catastrophe. The news was reported on April 28 in a five-sentence report by Tass, the official Soviet news agency.

Gorbachev announces details on national television

But the political situation in the Soviet Union in 1986 had made some strides beyond the Cold War policies prevalent a decade earlier. On May 14, Soviet leader Mikhail Gorbachev, in keeping with his new policy, *glasnost* (or openness), spoke on national television and described in detail all that was then known about the explosions. Still, it would take many months before all the details of the accident were unraveled.

Reactor crew conducts unauthorized experiment

The key event leading to the disaster was apparently an unauthorized experiment carried out by the plant crew. Operators wanted to know what would happen if there were a power outage and steam stopped flowing to the turbines. Would the kinetic energy of the spinning turbine blades be sufficient to maintain the cooling pumps until the emergency diesel generators turned on, they wondered. The way to find out, the crew reasoned, was to carry out a controlled test of this situation. What transpired was a series of six major mistakes by workers. Any one error by itself might not have been fatal, but the combination of all six proved cataclysmic.

Crew disables the reactor

The most serious debacle was the crew's decision to disable the reactor's emergency coolant system. At the outset of the test, the reactor began to lose power. Because the test could only be continued if the reactor remained in operation, the crew disabled the coolant system. (Remember that the reactor's design meant that whenever it lost coolant water, its fission rate—or power level—increased.) However, the reactor continued to lose power, so the crew removed all the control rods from the core. This move dramatically increased the reaction rate of the reactor. Almost instantly the fission rate was cause for concern.

Crew unable to reinsert control rods

As power in the core began to increase, the crew attempted to reinsert the control rods manually. But the channels into which the rods were supposed to fit had deformed because of heat in the core. The rods did not drop properly, and power release in the reactor went out of control.

Two cataclysmic explosions

As steam vented from the reactors, water levels dropped dramatically in the core. This loss of water in turn increased power output from fission reactions. In less than one second, power output from the core increased a hundredfold. As temperatures increased to more than 5,000° Celsius, parts of the core melted. Molten metal then reacted with the remaining coolant water and produced hydrogen gas, which blew off the top of the reactor. This was the first of the two explosions.

Details of the second explosion are less clear. Some authorities believe that it was largely a chemical and physical phenomenon, like the first explosion. They believe that it would take no more than a few seconds for the hydrogen gas initially produced to be ignited by heat released during the meltdown.

Was the second explosion a nuclear explosion?

Other experts believe, however, that the second explosion may have been a pure nuclear reaction. Theoretically, an RBMK can explode like a nuclear (atomic) bomb, unlike most reactors in use today. Some scientists believe that parts of the molten core may have achieved critical mass during meltdown. Then a true bomb-like explosion could have occurred, accounting for the second explosion.

Impact

The Chernobyl accident has had both short-term and long-term effects on the local area, on the world as a whole, and on the further development of the nuclear power industry. In addition to the 31 people who died immediately or within two weeks of the accident, another 299 were injured. About 135,000 residents were evacuated from the area within 18 miles of the damaged plant, and later another 200,000 from other areas. But these numbers do not begin to reflect the magnitude of the damage to human health over the next generation and beyond.

Plants and animals in the immediate area of Chernobyl and downwind of the plant were heavily contaminated by fallout, or radioactive particles in the air. Crops could not be harvested and most farm animals were destroyed to prevent their use as food. Almost a decade after the accident, levels of radiation are still so high in some areas that no native food can be grown or consumed. People survive on food that is shipped in from safe areas.

Long-term nuclear contamination of the soil

The chief contaminants remaining in the soil are cesium 137 and strontium 90. These two radioactive isotopes have half-lives of 30 and 28 years, respectively. That means that they will constitute a hazard for many more decades if they are not removed. Decontamination and removal of contaminated topsoil has not been progressing very rapidly, however. The result is that many residents of the area still face a constant and serious health risk from radioactive isotopes in the environment.

Fallout contaminates the international food supply

The fallout that covered the western Soviet Union spread to parts of Europe, causing concern about food supplies there also. On May 7, 1986, for example, Canadian customs officials announced that vegetables arriving from Italy were contaminated with radioactive iodine 31. At about the same time, member states of the European Community banned all fresh meat produced in Eastern Europe. As far away as Lapland, reindeer meat was so contaminated with radiation that it was declared unfit for human consumption.

Reactor will emit radiation for hundreds of years

And what of the reactor itself? Soviet engineers employed a wide array of techniques to put out the flames in the reactor, to cool it down, and to cover up the damaged facility. One of the first steps, for example, was to pump liquid nitrogen into the core to cool it down and to put out fires. Next, thousands of tons of sand, clay, lead, and boron were dumped on top of the plant from helicopters. These materials absorbed neutrons and helped put out fires. In the surrounding area, dikes were built to contain contaminated water, and several inches of soil were removed and transferred to a storage area.

Reactor's first tomb leaks

Eventually, workers began to build a huge sarcophagus over the damaged plant. The steel-and-concrete tomb is designed to isolate the ruins of Unit 4 for the hundreds of years during which it will continue to emit dangerously high levels of radiation. However, by 1992 the first shell cracked and began to leak radioactive material into the environment. The Ukrainian government found it necessary to develop a second shell, stronger than the first, to be built on top of the original sarcophagus.

Through the Chernobyl catastrophe, scientists hope to learn new information and increase safety in the future. At first, officials of the former Soviet Union were reluctant to change reactor design of their RBMK plants, since they felt that operator errors, and not plant problems, had caused the accident. Eventually, however, they adopted a number of modifications to make the reactors safer. For example, control rod design was altered so rods can never be removed completely from the reactor core.

How the Explosion Influences History

Outside the Soviet Union, however, scientists already knew what the Soviets were just finding out. Most non-Soviets had long been leery of the RBMK design and had rejected its use in their own countries. As one American nuclear expert said, "Most of the lessons from Chernobyl have been learned already and applied in the United States."

Another legacy of the Chernobyl accident is its impact on opinion worldwide about nuclear power, which varied from country to country. In the United States, where memories of the 1979 Three Mile Island accident had not yet faded, Chernobyl merely confirmed the fears that many had about the use of nuclear power. In France, Japan, Belgium, and other nations that depend heavily on nuclear power, the Chernobyl accident had hardly any impact. And in a few nations, such as Great Britain, the accident for the first time opened up a heated debate about the role of nuclear power as a source of energy.

Where to Learn More

Barringer, F. "Chernobyl: Five Years Later, the Danger Persists." *New York Times Magazine* (April 14, 1991): 28.

———."Four Years Later, Soviets Reveal Wider Scope to Chernobyl Horror." *New York Times* (April 28, 1990): A1.

Cooke, S. "Human Failures Led to Chernobyl." *Engineering News-Record* (August 28, 1986): 10–11.

Diamond, S. "Design Flaws, Known to Moscow, Called Major Factor at Chernobyl." *New York Times* (August 26, 1986): A1.

———."Moscow Now Sees Chernobyl's Peril Lasting for Years: Big Area Stricken." *New York Times* (August 22, 1986): A1.

Flavin, C. *Reassessing Nuclear Power: The Fallout from Chernobyl.* New York: Worldwatch Institute, 1987.

Jackson, J. O. "Nuclear Time Bombs." *Time* (December 7, 1992): 44–45.

Medvedev, G. *The Truth about Chernobyl.* Trans. Evelyn Rossiter. New York: Basic Books, 1991.

Moore, T., and D. Dietrich. "Chernobyl and Its Legacy." *EPRI Journal* (June 1987): 4–21.

Reinhardt, W. "Soviet Reactor Was Flawed." *Engineering News Record* (May 22, 1986): 16

Seneviratne, G. "Soviets Alter Nuclear Designs." *Engineering News Record* (September 4, 1986): 10.

Shaherhak, T. *Chernobyl: A Documentary Story.* Trans. Tan Press. New York: St. Martin's Press, 1989.

Sweet, W. "Chernobyl: What Really Happened." *Technology Review* (July 1989): 42–52.

Toman, B. "Disaster's Impact on Health Won't Be Known for Years." *Wall Street Journal* (April 23, 1987): 28.

U. S. Council for Energy Awareness. *Chernobyl Briefing Book.* Washington, DC: Government Printing Office, March 26, 1991.

U. S. Department of Energy. *Health and Environmental Consequences of the Chernobyl Nuclear Power Plant Accident.* Washington, DC: Government Printing Office, 1987.

U. S. Nuclear Regulatory Commission. *Report on the Accident at the Chernobyl Nuclear Power Station.* Washington, DC: Nuclear Regulatory Commission, 1987.

Van Der Pligt, J. "Chernobyl: Four Years Later; Attitudes, Risk Management, and Communication." *Journal of Environmental Psychology* (June 1, 1990): 91–99.

VII

Chemical and
Environmental Disasters

DDT insecticide contamination

Global
1939–

Background

DDT is the common name for the chemical compound dichloro-diphenyl trichloroethane. This compound was widely used as an insecticide from World War II until the early 1970s. It was hailed for its ability to reduce incidence of insect-borne diseases. On the negative side, however, its long-lasting toxic effects entered the entire food chain.

In addition to affecting plant, aquatic, bird, and animal life, DDT contamination extends to human beings as well. A recent study links DDT to breast cancer in women. The publication of Rachel Carson's *Silent Spring* in 1962 was primarily responsible for raising public consciousness in the United States about the ecological threat posed by DDT and other pesticides. Civic activism and lawsuits eventually led to formation of the U.S. Environmental Protection Agency (EPA) in 1970 and the ban on DDT use in the United States. The pesticide is still manufactured, however, and used in many countries, including Mexico.

Contamination of the ecological system results from excessive DDT use, exposing the destructive capabilities of pesticides.

DDT's blessing: Insect-borne human diseases are controlled

The chemical name for DDT is 2,2-bis (p-chlorophenyl)-1,1,1-trichloroethane. The compound has the empirical formula of $C_{14}H_9Cl_5$. DDT is a colorless crystalline powder that is slightly aromatic. It is insoluble in water, moderately soluble in alcohols, and highly soluble in some oils and fats. Its melting point is 109°C. DDT was first synthesized in 1874 by German chemist Othmar Zeidler. In 1939 Swiss scientist Paul Müller discovered its insecticidal properties. His discovery of DDT's effective-

Spraying oranges for aphids and red spiders in Southern California.

ness in controlling insect-borne human diseases earned him the Nobel Prize in medicine in 1946.

DDT was used widely and successfully during World War II to eliminate insect populations carrying typhus and malaria. It was also useful in preventing yellow fever, river blindness, elephantiasis, and bubonic plague. After World War II, DDT was primarily exploited for agricultural purposes. Until DDT became available, farmers and gardeners used arsenicals (compounds containing arsenic) to control pests. Arsenicals were very costly to manufacture, and DDT, which acts as a nerve gas on insects, was both cheaper and more effective. DDT was used extensively against beetles, moths, lice, butterflies, flies, and mosquitoes. By 1959 DDT manufacture rose to 124 million pounds annually for disease control, agriculture, gardening, and animal husbandry. It has been used throughout the world but most heavily in the tropical zones.

Details of the Controversy

The publication of Rachel Carson's *Silent Spring* in 1962 was largely responsible for alerting the public to the widespread ecological threat posed by DDT. The pesticide has proven to be a significant threat to the entire ecological system. It has contaminated the habitat and food supplies of large populations of plant, marine, bird, and animal life. In humans, where DDT's low solubility in water and high solubility in fat result in great bioconcentration, the primary danger lies in DDT's chronic effect (its effect over time).

DDT's curse: Its long-lasting contamination

Rachel Carson, born in 1907, attended Pennsylvania State College for Women and later Johns Hopkins University, where she earned a master's degree in biology. She taught at Johns Hopkins and the University of Maryland until 1936, when she took a position with the U.S. Bureau of Fisheries (which later became the U.S. Fish and Wildlife Service). A respected scientist and author, Carson wrote two books about the ocean and shore life before she published *Silent Spring*.

The theme of *Silent Spring* is the interdependence of life on earth. It stresses that polluting the ecosphere with pesticides or other toxic compounds results in unintended and unforeseeable consequences. The book blamed reduction in bird, fish, and certain mammal populations on indis-

Since the banning of DDT, other chlorinated compounds and organo-phosphates have been developed. Many of these compounds are much more toxic than DDT but lose their toxicity quickly and break down into relatively harmless products.

criminate DDT use. Carson did not denounce all use of pesticides—she advocated their selective use based on understanding their side effects.

Rachel Carson decries indiscriminate pesticide use in *Silent Spring*

Silent Spring begins with "A Fable for Tomorrow," which describes a village with no birds, fish, or wild animals. Another chapter documents the toxic effects of pesticides, especially DDT, on organisms. In a later chapter, "And No Birds Sing," Carson describes the spraying of elm trees with DDT to control Dutch elm disease. Robins feed on the earthworms under the trees, which have ingested the poison. The robins become sterile or die from the pesticide, and their population rapidly diminishes. The book's last chapter, "The Other Road," discusses alternative pest control methods, such as predator insects, bacteria, sterilants, crop rotation, and ecological biodiversity (allowing many plants and animals to interact).

Naturalists in the United States had noticed a reduction in local bird populations, but most did not suspect the cause until *Silent Spring* was published. Concerned citizens demanded that the government reevaluate and control pesticide use.

Impact

In 1962, the year *Silent Spring* was published, President John F. Kennedy, Secretary of Agriculture Orville Freeman, and Secretary of the Interior Stewart Udall turned their attention to ecological hazards posed by pesticide use. They intended to weigh the benefits in agricultural productivity against large-scale environmental contamination. Many members of the departments of agriculture and interior, with close alliance to pesticide manufacturers, resisted attempts to reevaluate or restrict DDT and other pesticides. The manufacturers aggressively attacked Carson and other supporters as "fanatics." Eventually, however, the ecological threat posed by DDT was widely understood.

EPA created to oversee agricultural chemicals

Congress amended the Federal Insecticide, Fungicide, and Rodenticide Act (FIFRA) in 1964, tightening label guidelines and requiring that safety information be provided. In 1970 President Richard Nixon created the Environmental Protection Agency (EPA), consolidating the major programs and agencies that dealt with environmental pollution. The EPA reg-

A helicopter with spray booms delivers a dusting of insecticide in the Cranston, Rhode Island, area to combat gypsy moth caterpillars (May 15, 1982).

ulates the introduction of new agricultural chemicals (and other compounds) into the marketplace and establishes safety guidelines and testing procedures. The state of the nation's ecological awareness is far from unanimous, however. The U.S. Department of Agriculture (USDA) and the EPA continue to disagree over which parameters are appropriate when conducting pesticide risk/benefit analysis.

Despite the slow response of the federal government, citizens used the judicial system to halt DDT use in many local and state jurisdictions; Michigan and Arizona, for example, banned DDT in 1968. Federal regulations began phasing DDT out in 1969, a year before the EPA was established. EPA administrator W. D. Ruckelshaus announced a ban on almost all uses of DDT at the end of 1972.

DDT banned from use in United States

Since the banning of DDT, other chlorinated compounds and organophosphates have been developed. Many of these compounds are much more toxic than DDT but lose their toxicity quickly and break down into relatively harmless products. While farmers applied pounds of DDT per acre, only grams per acre of the newer pesticides are required. Moreover, the newer compounds are often designed to destroy only specific insects or weeds.

Nonetheless, many pesticide problems persist. Based on scientific evidence available to her in the late 1950s and early 1960s, Carson predicted in *Silent Spring* that insects would become increasingly resistant to DDT and other pesticides. This prediction has proved accurate. In the 1930s, before DDT, American farmers lost about one third of their crops to insects, weeds, and disease. In the 1990s farmers still lose the same proportion of their crops, despite spending over $4 billion on pesticides. Now, however, the entire ecosystem contains toxic compounds that may affect future organisms in unknown ways.

Some contaminated people show no effects; others have cancer, behavior problems

The long-term effects of DDT on humans are not fully understood. International research shows that DDT is found in all human blood and fat. High levels are found in the milk of mothers in the tropics. There are instances of people being exposed to relatively large amounts of DDT without exhibiting obvious symptoms of poisoning, as in the small city of Triana, Alabama, which is downstream from the U.S. Army's Redstone Arsenal. Olin Chemical Co. manufactured DDT at this site from 1947 until 1970, pouring tons of DDT-laden wastewater into Indian Creek, which runs through Triana. The local fish and waterbird populations diminished significantly. However, Triana residents, who ate the Indian Creek fish and accumulated high levels of DDT and other chlorinated compounds, exhibit no noticeable effects.

Other evidence suggests a link between high DDT levels and behavior problems in children, increased suicides among older people, and pancreatic cancer or increased chances of pancreatic cancer. A 1993 study by Mary Wolff of Mount Sinai School of Medicine in New York City ties DDT levels in women to breast cancer. Whereas an increase in breast cancer was known to have coincided with the widespread application of DDT, Wolff's study, published in the *Journal of the National Cancer Institute,* is the first to establish a statistical link.

Where to Learn More

Carson, Rachel. *Silent Spring.* Boston: Houghton, 1962.

Garabrant, D. H., and others. "DDT and Related Compounds and Risk of Pancreatic Cancer." *Journal of the National Cancer Institute* (1992): 764–771.

Graham, Frank, Jr. *Since Silent Spring.* Boston: Houghton, 1970.

"The Joy Ride Is Over: Farmers Are Discovering that Pesticides Increasingly Don't Kill Pests." *U.S. News & World Report* (September 14, 1992): 73–74.

Lewis, Richard L., ed. *Sax's Dangerous Properties of Industrial Materials.* 8th ed. New York: Van Nostrand, 1992.

Silent Spring Revisited. Washington, DC: American Chemical Society, 1987.

Sittig, Marshall. *Handbook of Toxic and Hazardous Chemicals and Carcinogens.* 2nd ed. Park Ridge, NJ: Noyes, 1985.

Stranahan, Susan Q. "The Town that Ate Poison." *National Wildlife* (June/July 1984): 16–19.

Whorton, James. *Before Silent Spring: Pesticides and Public Health in Pre-DDT America.* Princeton, NJ: Princeton University Press, 1974.

Love Canal toxic waste site

Niagara Falls, New York
1942–80

A Niagara Falls, New York, neighborhood built over a toxic waste dump is abandoned in 1980 when it is declared a threat to public health.

Background

Love Canal is the first and probably best known of the nation's toxic waste sites. The name has come to symbolize chemical contamination of areas and neighborhoods.

Between 1942 and 1953 the Olin Corporation and the Hooker Chemical Corporation buried over 20,000 tons of deadly chemical wastes, including dioxin, in Love Canal, located near Niagara Falls, New York. Many of the compounds dumped there are known to cause cancer, miscarriages, birth defects, and other illnesses and disorders.

Hooker Chemical donates land to school board for $1

In 1953 Hooker Chemical donated the land to the local board of education for a token payment of one dollar. If it did not clearly warn of the dangerous nature of the chemicals buried there, neither did the buyer beware. Homes, playgrounds, and a school were built on the land. In 1976, after years of unusually heavy rains raised the water table and flooded basements, problems began to appear in the neighborhood. Homes reeked of chemicals; children and pets sustained chemical burns on their feet and hands, and some pets even died, as did trees, flowers, and vegetables.

Health problems begin to occur in mid-1970s

People in the neighborhood reported suffering serious and unexplainable illnesses. They had higher-than-normal rates of cancer, miscarriages, and deformities in newborns.

Section of the fence surrounding Love Canal where a residential community was built over a dumpsite containing more than 20,000 tons of deadly chemical wastes.

Alarmed and frustrated by the lack of action on the part of local, state, and federal authorities, Lois Gibbs, a 27-year-old housewife, took matters into her own hands in spring 1978. She organized her neighbors into the

Love Canal Homeowners' Association and worked for two and a half years to have the government relocate them to another area. Gibbs's work turned Love Canal into a household name across the country. It also helped focus nationwide attention on the problem of toxic waste disposal, which in December 1980 resulted in passage of the Environmental Emergency Response Act, known as the Superfund Law.

Details of the Disaster

With increasing numbers of people experiencing health problems in Love Canal, some residents began to suspect that the community might not be safe for human habitation. Local and national media gave extensive coverage to the neighborhood activists, and they brought pressure to bear on government officials. In August 1978 the New York State health commissioner, Dr. Robert P. Whalen, recommended that pregnant women and children under the age of two be evacuated, saying that there was "growing evidence . . . of subacute and chronic health hazards, as well as spontaneous abortions and congenital malformations" at the site. When the state tested the air, water, soil, and homes for toxic chemicals later that month, over 80 different compounds were found, many of which were thought to be capable of causing cancer. Chemical pollution of the air was measured at 250 to 5,000 times the levels considered safe.

High miscarriage rates in neighborhood

The 1978 study found an unusually high miscarriage rate of 29.4 percent in the neighborhood, with 5 of the 24 children born in the area listed as having birth defects. State health officials estimated that women in the area had a 50 percent higher-than-normal risk of miscarriage. Another report found that in 1979 only 2 of 17 pregnant women in Love Canal gave birth to normal children. Four had miscarriages, 2 had stillbirths, and 9 had babies born with defects.

Epidemiological studies of the affected population revealed an alarming pattern of illness among exposed residents. For example, on 96th Street, where 15 homes were located, 8 people developed cancer in the 12-year period between 1968 and 1980: 6 women had cancerous breasts removed; 1 man contracted bladder cancer; and another developed throat cancer. In addition, a 7-year-old boy experienced convulsions and died of kidney failure, and a pet dog had to be destroyed after developing cancerous tumors.

Drums containing contaminated sediment from Love Canal. The total cost for the cleanup has been estimated at $250 million.

Congress investigates

Some of the most alarming health data were gathered by Dr. Beverly Paigen of the Roswell Cancer Institute, who found a much higher incidence of illness among people who lived in houses that were located above moist ground or wet areas—those most prone to contamination by rising ground water. On March 21, 1979, testifying before the House Subcommittee on Oversight and Investigations, she described the tragic history of several families who lived in one such house located directly over liquid wastes that were seeping out of the ground. Among the four families, there existed three cases of nervous breakdown, three hysterectomies due to uterine bleeding, cancer, or both, and numerous cases of epilepsy, asthma, and bronchitis.

Dr. Paigen also found a significant excess of childhood disorders among youngsters born to residents of wet areas, including 9 instances of birth defects among the 16 children born in such areas between 1974 and 1978. She determined that the overall incidence of birth defects was 20 percent; the miscarriage rate was estimated at 25 percent, compared with just 8.5 percent for women moving into the area.

She also reported that 11 out of 13 hyperactive children lived in wet areas, and that 380 percent more asthma occurred there than in the "dry" areas of Love Canal. The incidence of urinary disease and convulsive disorders was almost triple that of dry areas, and the rates of suicides and nervous breakdowns almost quadruple. Significantly, Dr. Paigen found that Love Canal residents suffering from illnesses ranging from migraines to pneumonia to severe depression reported marked health improvements when they moved out of the area and away from the contamination.

Love Canal called "a grave and imminent peril"

Eventually, New York State authorities termed the area "a grave and imminent peril" to the health of those living nearby. Several hundred families were moved out of the neighborhood, and the others were advised to leave; the school was closed, and a barbed wire fence was placed around it.

In May 1980 further testing revealed high levels of genetic damage among neighborhood residents, resulting in an additional 710 families being evacuated at a cost estimated to run between $3 million and $60 million. In October 1980 President Jimmy Carter declared the neighborhood a disaster area. In the end, some 60 families decided to remain in their homes and reject the government's offer to buy them out. The total cost for the cleanup has been estimated at $250 million.

State begins allowing rehabitation of Love Canal

Twelve years after the neighborhood was abandoned, the state of New York approved plans to allow families to move back into some parts of the area, and homes were permitted to be sold.

Impact

The major impact of the massive publicity generated by Love Canal was to draw national attention to the dangers of toxic chemicals and hazardous waste, and to pressure Congress and the White House to pass laws to address the problems caused by such pollution. Hooker faced public outrage when information came out that it tried to disclaim its liability when it gave the property to the school board.

Hooker silent about dangers, and so is school board

When Hooker deeded the land to the Niagara Falls Board of Education for one dollar, the company was quiet about the lethal nature of the chemicals it had buried there. The legal document transferring the property disclaims liability for any deaths or injuries that might occur on the land and specifies that the school board would assume responsibility for any claims that might result from exposure to the buried chemicals. If Hooker tried to protect itself by quietly inserting disclaimers into the deed, the school board raised no questions about why Hooker felt it necessary to be concerned about liability.

Even when a neighborhood began to be developed in the area, no one raised warnings about the dangers there. On June 18, 1958, a company memo observed that "the entire area is being used as a playground," and that "3 or 4 children had been burned by material at the old Love Canal property." Ten years later, roadworkers building a highway near Love Canal uncovered leaking drums full of toxic chemicals. Hooker analyzed them and discussed them—in a March 21, 1968, internal company memo.

Industry called residents hypochondriacs

The chemical industry downplayed the threat to public health posed by Love Canal, as when the late Armand Hammer, then chairman of Hooker's parent company, Occidental Petroleum, said on the October 14, 1979, edition of NBC's "Meet the Press" that the Love Canal problem had "been blown up out of context." When Hammer chaired Occidental's annual stockholder meeting in May 1980, the company rejected a stockholder resolution calling on the firm to adopt policies designed to prevent future Love Canals. And during a July 2, 1980, PBS television interview on the *MacNeil/Lehrer News Hour*, vice president of the Chemical Manufacturers Association Geraldine Cox intimated that Love Canal residents were comparable to hypochondriacs with imaginary or exaggerated illnesses.

By 1980 information came out that there were thousands of other toxic waste dumps—potential Love Canals—scattered across the country. The U.S. Environmental Protection Agency (EPA) put the number of hazardous waste disposal sites in the United States in the 32,000 to 50,000 range, with 1,200 to 2,000 of those possibly posing "significant risks to human health or the environment." And new chemical waste sites were being discovered at a rate of 200 a month.

EPA estimates 90 percent of waste disposal is illegal

The EPA estimated that only about 10 percent of the 150 million tons of hazardous wastes being generated each year were disposed of in a safe and legal manner. The 90 percent remaining was being dumped illegally or disposed of in a way that posed a potential threat to humans or the environment. The agency called the situation "the most serious environmental problem in the U.S. today." (More recent estimates put the amount of hazardous waste being produced each year at 300 million tons—roughly a ton for every man, woman, and child in the country.)

On May 16, 1979, Assistant Attorney General James Moorman testified before the House Subcommittee on Oversight and Investigations that toxic waste dumping was the "first or second most serious environmental problem in the country." He pointed to a lack of effective environmental regulations and a lack of enforcement of existing antidumping laws.

Superfund Law passed, December 11, 1980

With public concern high and because of the uproar over Love Canal, laws designed to protect the public from toxic chemicals were passed or strengthened. In November 1980 a provision of the Resource Conservation and Recovery Act (RCRA) went into effect, requiring that toxic wastes be tracked "from cradle to grave." And on December 11, 1980, President Carter signed into law the Environmental Emergency Response Act—known as the Superfund Law—that created a fund to pay for the cleanup of hazardous waste sites and makes owners and operators of waste disposal sites, as well as producers and transporters of hazardous materials, liable for cleanup costs. These laws have made substantial reductions in the improper disposal of dangerous chemicals, but the task—which never seems to catch up with waste generation—is overwhelming. At the beginning of 1993, only 149 of 1,256 priority Superfund sites had been cleaned up.

Laws lead the way to recycling

Despite the waxing and waning of attention to environmental concerns of different presidential administrations, the new laws and regulations have eliminated many abuses. They have also led the chemical industry to begin recycling its waste—the industry reported recycling half of the 38 billion pounds it produced in 1991. Nevertheless, toxic waste disposal remains a critical problem. In May 1993 the Agency for Toxic Substances and Disease Registry estimated that:

• Over 64,000 areas in the United States contaminated by hazardous wastes require cleanup.

- Some 41 million Americans live within four miles of a hazardous waste site and may be at some risk.
- At least 250,000 people have been affected by "acute release events," such as waste releases, spills, or accidents.

Cleanup may cost $1 trillion and take a half century

It may prove impossible to completely clean up the nation's dumpsites at any price. The anticipated costs for steps that must be taken in the next few years are staggering. One EPA study estimated that it will cost over $44 billion just to clean up the most dangerous sites, with the public having to pick up half the tab. Other projections put the figure for addressing Superfund waste sites at over $1 trillion—and a timeline of half a century to accomplish it.

Where to Learn More

Blumenthal, R. "Fight to Curb 'Love Canals.'" *New York Times* (June 30, 1980).

Brown, M. H. "Love Canal Revisited." *Amicus Journal* (summer 1988): 37–44.

———. "A Toxic Ghost Town: Ten Years Later, Scientists Are Still Assessing the Damage from Love Canal." *Atlantic* (July 1989): 23–26.

"Fleeing Love Canal." *Newsweek* (June 2, 1980): 56.

Gibbs, Lois. *Love Canal: My Story.* Albany, NY: State University of New York Press, 1982.

Kadlecek, M. "Love Canal: Ten Years Later." *Conservationist* (November–December 1988): 40–43.

Regenstein, L. G. *Cleaning up America the Poisoned.* Herndon, VA: Acropolis Books, 1993, pp. 136–145.

Minamata Bay mercury poisoning

Minamata Bay, Kyushu, Japan
1955–

Industrial disposal of mercury wastewater causes Minamata disease. Thousands of Japanese citizens sue the Chisso Company and their local and federal governments.

Background

At least 2,900 Minamata Bay residents succumbed to organic mercury poisoning since 1953. The organic mercury was contained in and converted from the industrial waste of the Chisso Company, founded early in the century on Kyushu, the southernmost of Japan's four main islands. By midcentury the firm produced huge quantities of acetaldehyde and vinyl chloride, which generated mercury-laden waste responsible for the disease outbreak. Although the business admitted as far back as 1925 that its industrial waste harmed the local fishing industry, federal policies allowed it to continue dumping untreated waste into Minamata Bay. Fish and shellfish absorbed the organic mercury. Animals and humans ate the fish and shellfish, thus ingesting the mercury. The chronic effect of mercury ingestion is extremely disabling and often fatal in adults. It also produces congenital deformities in the unborn.

Minamata Bay supported large fishing industry

The city, river, and bay all named Minamata make up the Minamata Bay area in Kumamoto Prefecture, on the west coast of Kyushu island. Minamata has a population of approximately 37,000 and is surrounded by villages of fishermen and farmers. Kyushu's irregular coastline provides many ports, and to the west are the Amakusa islands, which guard Minamata Bay from the Shiranui Sea. Thus protected, Minamata Bay has always been known for its abundance of fish.

In mid-1900 Sogi Electric Company built an industrial plant in Minamata. The company shortly thereafter underwent two name changes, first

Mercury discharged from a local chemical plant contaminated fish and shellfish eaten by the residents of Minamata, many of whom contracted mercury poisoning as a result. Here, a victim of Minamata disease, or mercury poisoning, is being bathed by a relative.

to Japan Nitrogen Fertilizer Company (Nippon Chisso Hiryo Kabushiki Kaisha), then to New Japan Nitrogen Company. It has long been popularly known as the Chisso Company ("chisso" is Japanese for nitrogen).

Chisso Company very successful in 1950s and 1960s

The Chisso plant began producing acetaldehyde (C_2H_4O) in 1932. This compound is important in the manufacturing of plastics, pharmaceuticals, photographic chemicals, and perfumes. Other products Chisso manufactured also required mercury, especially vinyl chloride. During the 1950s and 1960s the plant became one of the largest production facilities for these compounds.

Law permits wastewater dumping—for the sake of growth

Mercury's role in the manufacture of Chisso's products is that of a catalyst. A catalyst aids the production process by causing specific chemical reactions to occur, but the catalyst usually does not become incorporated into the product. After the product is manufactured, the catalyst is present in the waste. If this manufacturing byproduct is recycled, the individual components, such as mercury, can sometimes can be reused, but recovering them is a very costly process. When recycling is not economical and especially when it is not required by law, the manufacturer disposes of the waste. The Chisso plant discharged its mercury and other industrial waste into Minamata Bay.

The form of mercury that the Chisso Company used in its production processes was inorganic mercury. Classified as a heavy metal, it is silver in color and liquid at room temperature, which is why it has been called quicksilver. This is the fluid found in fever thermometers.

Mercury is toxic to animals and humans

Humans have long used inorganic mercury—evidence shows that mercury was known to the ancient Chinese and Hindus, and it has been found in Egyptian tombs dating back to 1500 B.C. Mercury salts are useful medicinally as antiseptics. But it is toxic when, over time, it accumulates in living organisms.

Among the ill effects caused by inorganic mercury poisoning are tremors, loose teeth, headaches, fatigue, and minor psychological changes. Laboratories and hospitals take precautions to ensure that inorganic mercury does not come into contact with the skin and is not ingested.

The Chisso Company created an even greater hazard—organic mercury, which formed either directly from their manufacturing processes or resulted from microbial action on the inorganic mercury it dumped into the bay. Certain microbes found in lakes, rivers, or ocean sediments can convert inorganic mercury into organic mercury ("organic" indicates that the compound is based on the carbon molecule, which is the basis of living organisms). Water conditions—such as acidity, temperature, and the presence of other metals—appear to influence this reaction.

In just four years, Chisso dumps 200 tons of mercury

For every ton of acetaldehyde that Chisso produced, the company discharged from 500 to 1,000 grams of inorganic mercury in its wastewater. Experts estimate that between 1949 and 1953 more than 200 tons of the poison were discharged into Minamata Bay.

While it is not understood how organic mercury affects fish or shellfish, it does not flush out of the organism that ingests it; rather it accumulates as more is taken in. Then, as larger animals eat the smaller, the mercury from the smaller animal is deposited into the larger, and so on up the food chain. Eventually, as birds, larger animals, and humans eat the fish and shellfish, their bodies amass the poison until the level is high enough to cause significant health problems. The chronic (long-term) effect of organic mercury is very toxic to humans. Organic mercury concentrates in the brain, liver, kidneys, and in the human fetus. Testing hair, blood, and urine can reveal the level of mercury poisoning a person has acquired.

Fish in the bay tested at 50 ppm

In the United States, the FDA (the Food and Drug Administration) set standards for the highest amount of mercury contamination permitted in fish: when intended to be eaten, fish may not have more than 0.5 ppm (parts per million). Fish tested from Minamata Bay registered levels of organic mercury as high as 50 ppm. Measuring such small levels of organic mercury in fish is a complex process—and while procedures exist to detect organic mercury at levels as low as 0.5 ppb (parts per billion), few laboratories have the equipment to perform these tests.

In 1925, long before the chronic effects of organic mercury poisoning became evident in Minamata Bay area residents, Chisso admitted that its waste was harmful to the local fishing industry. It agreed to compensate the local Fishermen's Cooperative and paid out a token amount. Chisso made no effort to treat its wastewater, and the government did not require it.

Japanese government encourages industrial growth—at any cost

At that time the Japanese government was encouraging industrial growth throughout the country, and there was little concern for the potential effects of industrial pollution. Evidence exists showing that whenever poisoning outbreaks occurred that were due to industrial waste, Japanese industries together with the central government intimidated and suppressed the criticism and complaints.

Details of the Disaster

In the mid-1950s local health officials began seeing an unusual outbreak of a strange disease in the villages around Minamata. By 1956 the medical director of the Chisso Company hospital and the Kumamoto University School of Medicine agreed with the findings of the local health office. Investigations established that the disease, which became widely known as Minamata disease, had taken its first official victim in 1953.

In April 1956 Dr. Hajime Hosokawa, the medical director of the Chisso Hospital, began examining patients with symptoms of the disease. He reported to the Minamata Public Health Department that an unidentified disease had afflicted a number of people in the area. He was concerned that it was a contagious disease, perhaps a form of polio. The disease affected birds and mammals as well as humans, and there were many sick cats.

Chisso suppresses Dr. Hosokawa's research

Dr. Hosokawa discovered the actual cause of the disease in July 1959 by giving waste discharges to a cat (No. 400) and watching it develop the same symptoms as the humans and cats that were known to be ill. The Chisso Company stepped in and ordered Dr. Hosokawa to stop his experiments and to report nothing to health authorities.

Another agency implicates Chisso

Minamata public health officials turned to the Kumamoto University School of Medicine for assistance in August 1959. The medical school sent a research team to investigate and report directly to city officials. The team suspected that the symptoms resulted from the ingestion of heavy metals. They studied the effects of magnesium, selenium, thallium, and other heavy metals. By 1962 they were certain that organic mercury caused the symptoms of the disease and that the source of the organic mercury was the industrial wastewater from the Chisso plant.

People and animals who succumbed to Minamata disease had eaten raw fish and shellfish contaminated with organic mercury. In early stages victims experienced numbness of the lips and limbs, followed by constriction of vision. Muscular coordination then declined, and victims exhibited exaggerated reflexes and their walking became very unsteady. They lost their ability to speak, hear, and taste, and they were prone to violent behavior or depression. Advanced victims became stiff and bent, and their blood pressure and body temperature soared just before they died. Babies born to mothers with mercury poisoning were retarded, blind, and deaf, their limbs were deformed, and they were unable to walk or speak.

Impact

The Chisso Company was indicted and tried for causing Minamata disease by discharging its industrial waste into Minamata Bay. Plaintiffs sued Chisso for damages. The trial dragged on—from July 1969 until March 1973—when Chisso was finally convicted. The court ordered the manufacturing company to pay group damages and begin negotiations to pay damages to individuals.

Plaintiffs win awards

No one knows how many people were actually stricken with Minamata disease. The government officially recognized 2,900 victims as entitled to damage awards, but there were hundreds and perhaps thousands more. Beyond the 2,900 official victims were the plaintiffs who did not win damage awards, because their symptoms did not strictly match the official profile of the disease. There were also other Minamata victims who did not seek legal redress for fear of the stigma it would bring on their families.

The plaintiffs also took their prefecture and federal governments to court. They intended to force the government to recognize its complicity and to pay damages to the victims. In 1992, however, a Japanese federal district court determined that the federal government was not responsible for damages to the victims. This determination went counter to the precedent set in a 1907 decision in Kumamoto Prefecture court—that both the federal and prefectural governments were legally culpable. The victims of Minamata disease continue to press the suit in their courts.

Waste in Minamata Bay is being covered over

Measures to correct the pollution are also underway. Minamata Bay is being filled in to cover the mercury-laden industrial waste. There is a proposal to convert the area into an environmental park. Fishing has been banned from the bay, but contaminated fish and shellfish from just outside the bay are still caught, sold, and eaten.

The area is suffering economically

Minamata residents continue to feel the effects of the scandal. They claim they are stigmatized in their ability to find jobs and even spouses in other parts of Japan. The city of Minamata and Chisso Company are both suffering. Most local citizens sided with Chisso. They resented the legal actions taken by the fishermen against the company, because they feared that the lawsuits would hurt Chisso financially and therefore their own standard of living.

Where to Learn More

D'Itri, Patricia A., and Frank M. D'Itri. *Mercury Contamination: A Human Tragedy*. New York: John Wiley, 1977.

Ishimure, Michiko. *Paradise in the Sea of Sorrow*. Trans. Livia Monnet. Japan: Yamaguchi, 1990.

Kuznetsov, D. A. "Minamata Disease: What Is a Keystone of Its Molecular Mechanisms? A Biochemical Theory on the Nature of Methyl Mercury Neurotoxicity." *International Journal of Neuroscience* (July 1990): 1–51.

Mishima, Akio. *Bitter Sea: The Human Cost of Minamata Disease*. Trans. Richard L. Gage and Susan B. Murata. Japan: Kosei, 1992.

Sanger, David E. "Japan and the Mercury Poisoned Sea: A Reckoning that Won't Go Away." *New York Times* (January 16, 1991): A3.

Schlesinger, Jacob M. "Japan Government Cleared of Blame in Minamata Industrial Pollution Case." *Wall Street Journal* (February 10, 1992): A10.

Smith, W. Eugene, and Aileen Smith. *Minamata*. New York: Holt, 1975.

Tsuchiya, Kenzaburo. "The Discovery of the Causal Agent of Minamata Disease." *American Journal of Industrial Medicine* (1992): 275–280.

Agent Orange contamination

Background

Best known for its use during the Vietnam war, Agent Orange defoliated jungles extremely well. However, the herbicide is contaminated with a deadly compound, dioxin. Extensive use of Agent Orange produced long-range health effects in people exposed to it, both civilian and military populations. Years after the Vietnam conflict, a number of American military personnel are suffering various health problems that they claim are caused by their exposure to Agent Orange.

Extensive use of the defoliant Agent Orange during the Vietnam war causes a cluster of medical problems among both the Vietnamese population and American military personnel.

Agent Orange kills vegetation effectively

Agent Orange is a very effective herbicide used to clear heavy vegetation, especially in forests, jungles, and rights-of-way. Two herbicides—2,4-D and 2,4,5-T—combine in equal measure to make the Agent Orange formula. Agent Orange was first used commercially in forestry control as early as the 1930s. U.S. chemical manufacturers discovered the defoliating properties of these herbicides and marketed the chemicals domestically and for export. During the 1950s and 1960s New Brunswick, Canada, was the site of heavy Agent Orange spraying to control forests for industrial development. In the 1950s the British used compounds with the chemical mixture 2,4,5-T in Malaysia to clear communication routes.

Toward the end of World War II the U.S. military saw the usefulness of the herbicides during action in the Pacific. In 1959 the Crops Division at Fort Detrick, Maryland, initiated the first large-scale military defoliation effort. Their aerial application of Agent Orange to about four square

The herbicidal effects of Agent Orange in Vietnam were readily apparent. Above is an untreated mangrove forest.

miles of vegetation proved highly successful—military personnel believed they had discovered an effective defoliation tool.

South Vietnam government wants to use Agent Orange

Aware of these early experiments, the South Vietnamese government thought the herbicides would be useful against insurgent guerrilla forces. By 1960 South Vietnam requested the United States to conduct trials of the herbicides. President John F. Kennedy agreed on condition that the program be carefully and selectively controlled. Kennedy was anxious that the program not affect the food supply.

Secretary of State Dean Rusk was also concerned. Rusk warned that U.S. involvement in defoliating the jungles of South Vietnam could possibly make the United States a target of a "germ warfare" campaign initiated by Communist countries and echoed by some neutral countries. Rusk urged that the program be closely controlled and supervised directly by Washington.

Above is another forest as it appeared approximately two years after being sprayed with Agent Orange.

Military chiefs get to control spraying

The military, however, recognized the limitations of fighting in foreign territory with troops unaccustomed to jungle conditions. They argued that clearing the communication lines and opening up areas of visibility would enhance their opportunities for success. Admiral Harry Felt, then head of operations in the Pacific, and the military joint chiefs did not want to have to get Washington's specific approval for each strategic decision. They wanted control to plan and conduct future herbicide operations. Defense Secretary Robert S. McNamara agreed with the Pentagon, and Rusk was overruled. Spraying began in 1961.

After 1962, the spraying increased monthly

In the early stages, 15 varieties of chemicals were employed as herbicides, with different components and formulas. Many contained some amount of 2,4-D or 2,4,5-T. The "rainbow of destruction" colors included blue, purple, pink, and green. These early varieties were used in low quantities (a total of 281,000 gallons) and on a limited geographical area. With Project Ranch Hand—as the spraying missions conducted by the Air Force were called—Agent Orange became the most widely produced and dispensed defoliant in Vietnam, and after 1962 the spray missions increased monthly.

Details of the Herbicide Use

Military procedures for herbicide use were developed by the army's Chemical Operations Division, J-3, Military Assistance Command, Vietnam (MACV). The military promised to stay away from civilians or to resettle civilians and resupply food in any areas where herbicides destroyed the food supply, but the reality was that these promises were difficult to keep. The use of herbicides for crop destruction peaked in 1965, when 45 percent of the total spraying was designed to destroy crops.

Herbicide demand outstrips supply

Initially the aerial spraying took place near Saigon, but the geographical base eventually widened. During 1967 the military greatly expanded its orders for the herbicide, until it outstripped the ability of the manufacturers to produce sufficient amounts of it.

The Air Force and the joint chiefs of staff become directly involved at that point. They diverted all commercial production to the military. Any problems associated with production and procurement of the defoliants were handled by the Department of Defense (DOD). Commercial producers were encouraged to expand their facilities and build new plants. The DOD even made attractive offers to companies to manufacture herbicides. Working closely with the military, certain chemical companies sent technical advisers to Vietnam to instruct personnel on the methods and techniques necessary for effective use of the herbicides.

During the peak of the spraying, approximately 129 missions were flown per aircraft. Twenty-four UC-123B aircraft were used, averaging 39 flights per day. Trucks and helicopters also conducted spraying missions, backed up by such countries as Australia. C-123 cargo planes were used as well. Helicopters flew without cargo doors so that frequent ground fire could be returned. Because the enemy was hidden and protected by the thick, broad leaves, the dense jungle growth required two applications of spray: one for the upper layer of vegetation and a second for the lower.

Civilians and soldiers exposed to spray

The foliage was not the only form of life exposed to the herbicides. American troops were heavily exposed. Rotary blades would kick up gusts of spray, delivering a powerful dose onto the faces and bodies of the men inside the plane. Soldiers could inhale the fine misty spray or be completely drenched in a sudden and unexpected emergency dumping.

The chemicals penetrated skin and lungs. Military personnel were exposed as they lived on the sprayed bases and slept near empty drums. They drank water in areas where defoliation had occurred. They swam, washed in, or slogged through water in bomb craters and rivers. They ate food misted with spray. Empty herbicide drums were indiscriminately used and improperly stored. Volatile fumes from these drums damaged shade trees and affected anyone near them.

Those whose direct responsibilities required them to handle the herbicides were exposed on a consistent basis. In addition to the Ranch Hands, at least three other groups received consistently high exposure:

- Secondary support personnel. These included army pilots who may have been involved in helicopter spraying, along with navy and even marine pilots.
- Those who transported the herbicide. They moved the omnipresent 55-gallon drums to Saigon, to BienHoa, to DaNang. . . .
- Specialized mechanics, electricians, and technical personnel assigned to work on various aircraft. Many in this group were not specifically assigned to Operation Ranch Hand but worked in contaminated aircraft.

Military used more chemical than recommended

Agent Orange was used in Vietnam in undiluted form at the rate of three to four gallons per acre; 13.8 pounds of the chemical 2,4,5-T were added to 12 pounds of 2,4-D per acre, just about a 1:1 ratio. This concentration is 13.3 pounds per acre more than was recommended by the military's own manual—far more than was ever used domestically. Computer tapes now available show that some areas were sprayed as many as 25 times in just a few short months, dramatically increasing the exposure to anyone within those sprayed areas.

Up to 19 millions gallons used

Agent Orange was very effective. Evaluations show that the chemical killed and defoliated 90 to 95 percent of the treated vegetation. Thirty-six percent of all mangrove forest areas in South Vietnam were destroyed. Viet Cong tunnel openings, caves, and aboveground shelters were revealed to the aircraft after the defoliation.

From 1961 until 1971, between 17 million and 19 million gallons of the liquid acid were sprayed, covering at least 10 percent of the total surface area in Vietnam. (The Vietnamese government claims over 44 percent of the land was sprayed.) Sections of both Laos and Cambodia were also sprayed.

Agent Orange is contaminated by dioxin—"one of the most toxic materials known"

A serious problem with Agent Orange is that it is contaminated by the deadly chemical dioxin, or TCDD. Dow Chemical (one of the Agent Orange manufacturers) knew since 1937 that 2,4,5-T was contaminated by some unidentified toxic chemical. Dow identified the contaminant as dioxin by 1965.

Dow met with all the Agent Orange contractors to inform them about dioxin's toxicity. Dow revealed that chloracne, a skin eruption resembling acne, was an indicator that the whole body was poisoned. Reports from the 1965 meeting identify dioxin as "one of the most toxic materials known, causing not only skin lesions but also liver damage [and] fatigue." Dow personnel customarily took extreme precautions in handling the materials.

Agent Orange manufacturers keep dioxin contamination quiet

Dow and its fellow contractors recommended keeping the dioxin contamination low. This is possible if the manufacturing temperature is carefully controlled. Repeat exposure at even 1 ppm (part per million) was acknowledged to be a significant health hazard.

In Vietnam the dioxin levels varied from many parts per billion (ppb) to parts per million (ppm). The Environmental Protection Agency (EPA) evacuated Times Beach, Missouri, when tests revealed soil samples there with 2 ppb.

Impact

A cluster of medical problems experienced by American veterans of the Vietnam war were featured in a CBS television exposé about Agent

Orange in 1978. Veterans felt they were being given a runaround by the Veterans Administration (VA) and even accused the VA of engaging in a cover-up.

Meanwhile, the facts about dioxin continued to mount. Two federal agencies, the Agency for Toxic Substances and the Disease Registry Toxicological Profile for Dioxin, produced enough evidence to "conclude that dioxin deserves our greatest respect." According to a congressional research brief, "TCDD is classified as class 6, supertoxic in a range of 1 (practically non-toxic) to 6. TCDD may be the most toxic and potent teratogen [a substance that causes developmental malformations] known to man based on toxicological data in guinea pigs and rabbits." According to Dr. Barry Commoner, director of the Center for the Biology of Natural Systems at Washington University, the chemical is "so potent a killer that just three ounces of it placed in New York City's water supply could wipe out the populace."

Dioxin interferes with various systems of the body

Dioxin research has been conducted since the 1940s, initially on animals in laboratories, and indicates that the chemical compound appears capable of interfering with a number of physiological systems.

EPA scientists and a broad spectrum of researchers call dioxin the most "carcinogenic" compound ever studied. During the late 1980s and early 1990s, scientists confirmed its potential to cause rare forms of cancer in humans—especially soft tissue sarcoma and non-Hodgkins lymphoma.

List of toxic effects continues to grow

The EPA considers dioxin in food 1 million times more toxic than cadmium or arsenic. Recent EPA studies indicate that low-level exposure leads to immune system damage (a damaged immune system leaves the body less able to defend itself against hostile forces in the natural environment) and reproductive dysfunction. It appears to be most damaging to young animals exposed in the uterus. It also appears to affect behavior and learning ability, suggesting that it is neurotoxic.

Dioxin produces problems with the body's chemical systems year after year, functioning like a steroid hormone. Newer findings conducted on wildlife around the Great Lakes showed that dioxin has the capacity to feminize male chicks and rats and masculinize female chicks and rats. Testicle size is reduced, as is sperm count.

Veterans Administration agrees to pay for Agent Orange-caused illnesses

When the American Legion did its own studies on Vietnam veterans and reviewed the scientific literature, it concluded that at least 30 cancers and diseases are related to Agent Orange because of the dioxin contaminant. In 1993 the Institute of Medicine of the National Academy of Science announced that dioxin is now linked to Hodgkin's disease and a skin-blistering condition knows as porphyria cutanea tarda. The Veterans Administration acknowledged the impact of Agent Orange on human health, meaning that veterans will have their Agent Orange-related claims compensated.

Where to Learn More

Agent Orange Scientific Task Force, working with American Legion, Vietnam Veterans of America, and National Veterans Legal Services Project. "A Review of the Scientific Literature: Human Health Effects Associated with Exposure to Herbicides and/or Their Associated Contaminants; Chlorinated Dioxins: Agent Orange and the Vietnam Veteran." April 1990.

Gough, M. "Agent Orange: Exposure and Policy." *American Journal of Public Health* (March 1991): 289–290.

———. *Dioxin, Agent Orange: The Facts.* New York: Plenum, 1986.

Husar, R. B. *Biological Basis for Risk Assessment of Dioxins and Related Compounds.* Cold Spring Harbor, NY: Cold Spring Harbor Laboratory Press, 1991.

———. *The Health Risks of Dioxin.* Hearing before the Human Resources and Intergovernmental Relations Subcommittee on Government Operations, House of Representatives, 102d Congress, June 10, 1992. Washington, DC: Government Printing Office, 1993.

Schmidt, K. F. "Dioxin's Other Face: Portrait of an 'Environmental Hormone.'" *Science News* (January 11, 1992): 24–27.

Tschirley, F. "Dioxin." *Scientific American* (February 1986): 29–35.

Zumwalt, Admiral Elmo R., Jr. USN (Ret.). "Report to the Secretary of Veterans Affairs, The Hon. Edward J. Derwinski. From the Special Assistant: Agent Orange Issues." First Report, May 5, 1990.

Union Carbide toxic vapor leak

Bhopal, India
December 3, 1984

Background

On the morning of December 3, 1984, a poisonous gas cloud escaped from the Union Carbide chemical plant in Bhopal, India, and drifted over the city. Composed of methyl isocyanate gas, a compound from a pesticide produced at the plant, the cloud formed when water mistakenly entered the wrong tank. The gas increased the pressure inside the tank until it blew open a valve and released 50,000 pounds of deadly gas. Thousands of people died, and many more were badly injured. The horrible accident at Bhopal prompted harsh public scrutiny of chemical plants worldwide, and the long process of litigation over damage claims began in the courts.

Poisonous gas explosion at the Union Carbide plant in Bhopal, India, kills thousands and injures thousands more.

Union Carbide stored three times more MIC in its Bhopal tanks than other manufacturers

The Union Carbide site at Bhopal was first developed in 1969 as a mixing and packaging plant for pesticides imported from the United States. In 1980 the plant expanded into manufacturing Sevin and Temik. Both are "carbamate" pesticides (pesticides containing a salt or ester of carbamic acid). The formula for Sevin and Temik required large quantities of methyl isocyanate (MIC), a highly reactive, volatile, and toxic chemical compound. MIC was stored on-site in large underground storage tanks. These tanks each held approximately 15,000 gallons. Bhopal's tanks held three times the volume of the tanks used by other chemical manufacturers.

A toxic vapor leak at this Union Carbide plant in Bhopal, India, resulted in massive loss of life and widespread injury.

Methyl isocyanate is ten times more toxic than mustard gas

The toxicity of methyl isocyanate is well established, and the chemical industry and government regulatory agencies provide explicit warnings about MIC's handling and use. To limit the risk to human health, exposure to MIC is strictly controlled by the U.S. Occupational Safety and Health Act (OSHA). OSHA set the threshold limit value (TLV) for MIC exposure to 0.02 ppm (parts per million) in an eight-hour period. This is the maximum allowable amount a worker may be subjected to. Exposure beyond this limit is considered injurious to a worker's health. Two ppm of methyl isocyanate irritates the eyes, nose, and throat; at 21 ppm the irritation is tortuous. The threshold limit value of methyl isocyanate is one-tenth that of phosgene, or mustard gas, a chemical weapon used in World War I.

To handle and store methyl isocyanate, workers must take several safety precautions. They must protect stored MIC from overheating, from escaping into the atmosphere, and from being contaminated with other products, including water, with which it may react. MIC is also highly reactive with such common metals as iron, copper, tin, and zinc. Chemical workers often protect and cool MIC stored in bulk by "covering" it with a blanket of dry nitrogen. They prevent or slow reactivity of bulk-stored MIC by maintaining the temperature at 0°C or colder. Contaminating MIC with metal, water, or any other substance with which it reacts will create an extreme chemical reaction that is so serious it can escalate into a runaway reaction. Such an uncontrolled reaction, or chain reaction, creates heat that cannot be dissipated. The heat continues building until the container explodes.

Recordkeeping at Bhopal was lax

What exactly happened in the Bhopal explosion is still being debated. Most modern processing plants maintain extensive databases of records and operator logs. Electronic sensors and recording devices are installed and are linked to computers. Data collection and storage is thorough. Such modern plants also have computer models of cause-and-effect scenarios. In the event of accidents or chemical releases at these facilities, investigators have access to extensive records. Because recordkeeping systems in Bhopal, India, were far from ideal, investigators have been able to piece together only a most-probable-cause scenario.

Details of the Disaster

For six weeks before the accident, the Bhopal plant's MIC production unit had been shut down due to an oversupply of on-site stores of Sevin and Temik. As is usual during downtime, workers had been repairing, cleaning, and maintaining the facility. One of the standard maintenance duties was to clean the filters in the four lines that carried MIC from the storage tank into the processing unit.

These lines were connected to the pipeline of the relief valve vent header (RVVH), which carried toxic vapors from the MIC storage tanks to the vent gas scrubber (VGS) if pressure ever built up in the storage tanks. The VGS would "scrub" a vapor release with caustic soda. This would neutralize the MIC. The scrubbed gas would then be shunted to a knockout pot, a device for removing liquid droplets prior to flaring. The vapor release would be passed through the knockout pot, any droplets that might have formed would be removed, and then the remaining gas would be burned in a flare tower.

Filter-washing continues despite faulty drainage

To prevent water from leaking into the MIC tank through the RVVH piping during the washing process, a maintenance worker closed a valve. The worker believed that the filter pipelines were now isolated from the RVVH. However, the worker failed to insert a slip blind—a metal disc—into the valve to seal it, as is usually done when filters are washed. Also, two of the four bleeder valves that allow water to drain out of the lines were completely clogged; the other two were partially clogged. The washing process was temporarily halted when the drainage problems were noticed, but the filter-washing procedure soon resumed.

Leaky valves were common

An inspection conducted earlier in the year recorded that many valves in the plant were leaky. The filter isolation valve was no exception. With more water collecting than could be drained by the partially clogged filter washing system, water began flowing past the isolation valve and into the RVVH.

The pipes between the relief valve vent header and the MIC storage tank contained a number of closed valves designed to prevent water from flowing toward the storage tank. A one-way rupture disc was also installed. The rupture disc was designed to break if MIC vapor built up, releasing the vapor into the RVVH.

Plant installed a cheap, less reliable "backup" system

However, in May 1984 a jumper line had been installed between the RVVH and the process vent header (PVH). The PVH collected gases released when MIC moved from the storage tanks to the production unit. Any gases released from the tank would vent to the scrubber. If one of the header units were down for repairs, the jumper line allowed gases from the storage tank to vent to the scrubber through the headers of either the PVH or the RVVH. Installing a jumper line was the cheap but less dependable fix; the more expensive and more reliable solution would have been to install a backup line for the process vent headers.

The jumper line was opened on December 2 to allow repairs to take place on the process vent header. When water began backing up from the filter washing process, it easily flowed into the PVH instead of the RVVH. Once in the PVH, the water entered tank 610 through the leaky valves connecting the PVH to the storage tank.

The toxic fumes leaking from the plant were immediately deadly; people five miles away from the plant died from them. Eventually, thousands were evacuated from their homes.

Operators unable to pressurize tank or fix leaky valves

Operators knew the valves between the tank and the PVH were leaky because they had recently tried to pressurize tank 610 with liquid nitrogen. This was necessary to transfer MIC to a Sevin production unit, but they were unable to pressurize the tank. Despite repeated attempts for a week, operators were unable to locate and correct the faulty valve or valves.

Several other dangerous conditions contributed to the imminent Bhopal disaster:

- Tank 619 was supposed to be empty as a safety overflow, but it contained approximately 12 tons of MIC.

- Tank 610 was 75 to 87 percent full; Union Carbide safety standards specified that tanks should be no more than 50 to 60 percent filled.

- The MIC in tank 610 was contaminated by chloroform at a level much higher than Union Carbide's safety standards permitted. Chloroform reacts both with MIC and with the stainless steel walls of the storage tank.

- The temperature of the MIC in the storage tank was between 15° and 20°C. Standard safe storage procedures specify temperatures of 0° to 5°C, preferably lower.

Disabled safety systems useless as tank pressure increased

Operators on duty watched helplessly as the pressure in tank 610 climbed suddenly from 2 psi (pounds per square inch) to 30 psi to 55 psi. The fact that maintenance was underway further undermined the other safety systems.

The vent gas scrubber, on standby status, could not be turned on in time to neutralize any of the escaping gases. The flare tower was disabled because the pipe between the MIC tank and the vent gas scrubber had been removed. Also, operators were reluctant to move the reacting MIC into the overflow tank, since it was not completely empty. Trying to douse the leaking toxic vapors as they spewed out of a stack on the relief valve vent header didn't work, either. The stack was 120 feet high, but the spray reached only 100 feet.

Toxic vapor covers 15 square miles

For the next 90 minutes, approximately 27 tons of MIC vapor and 14 tons of reaction byproducts were sent into the atmosphere from tank 610, wreaking destruction for 15 square miles.

The white vapor—twice as heavy as air—filled the low-lying areas downwind of the plant. The narrow streets of the slums filled with people and animals gasping for breath. A neighborhood four miles from the plant sustained the greatest loss of life, but the deadly gas killed people five miles away.

Thousands die, hundreds of thousands suffer injuries

The Indian Supreme Court says officially that 3,000 people died from the accident. Other investigators believe the figure may be as high as 5,000. But medical personnel who directly treated the victims in Bhopal claim that as many as 12,000 died as a direct result of the leak. Some victims died immediately from either asphyxiation or suffocation caused by pulmonary edema (fluid in the lungs). Others died later from causes directly attributable to being exposed to the toxic vapor.

Those who survived the accident suffer from various symptoms, including acute respiratory distress, eye irritation, and problems of the circulatory, gastrointestinal, and central nervous systems. Long-term effects include chronic lesions of the eyes, permanent scarring of the lungs, and injuries to the liver, brain, heart, kidneys, and immune system. An epidemiological study of the yearly rate of spontaneous abortions and infant deaths in the years since the accident show the rate in Bhopal to be three to four times the regional rate.

An estimate of people injured by exposure to the vapor compiled by a commission for the Indian government included 30,000 with permanent injuries, 20,000 with temporary injuries, and 150,000 with minor injuries. However, victims' rights organizations believe the real numbers are actually much higher.

Impact

In the aftermath of the disaster, investigators inquired into the causes of the release and the potential for accidents at other processing facilities using and storing MIC. These inquiries examined several issues:

• Tank construction and monitoring

• Site safety warning and containment systems

• Potential for operator errors to cause failures

- Site configurations that undermine original safety features of the MIC storage tank and production unit
- Storage system safety maintenance and emergency response procedures

Investigators concluded that the Union Carbide India plant operated in an unsafe manner, providing the preconditions for a disaster to occur.

Union Carbide declares accident an act of sabotage

Indian prosecutors brought criminal negligence charges against the Indian and American management of Union Carbide. Some months after the accident Union Carbide claimed that an angry employee sabotaged the plant, deliberately allowing water into the MIC storage tank, but no one was ever formally charged.

After negotiating for several years with the government of Rajiv Gandhi, Union Carbide agreed to pay $470 million, and the government agreed to drop all criminal charges against the Union Carbide officials.

There were legal challenges to the amount of the settlement, which the Indian Supreme Court declared valid in 1991. The present Indian government, however, does not recognize the Supreme Court's decision and believes the original claim against Union Carbide, for $3.3 billion, is more reasonable. It also wants to pursue criminal charges against Carbide management. Meanwhile, Union Carbide India has paid an interim payment of $190 million, which the Indian government will distribute to survivors, pending the final outcome of the settlement.

Under the Bhopal Claims Act, the Indian government is currently giving survivors 200 rupees a month (about $10 U.S.). These payments will continue until the Bhopal case is settled—a process that will likely take several more years.

Where to Learn More

"Bhopal Tragedy's Health Effects." *Journal of the American Medical Association* (December 5, 1990): 2781.

Kurzman, D. *A Killing Wind: Inside Union Carbide and the Bhopal Disaster.* New York: McGraw-Hill, 1987.

Lepowski, W. "Indian Activists Press Bhopal Accident Issues in U.S. Visit." *Chemical & Engineering News* (May 13, 1991): 22–23.

———. "Lessons from Bhopal." *Chemical & Engineering News* (November 17, 1986): 39–45.

"Public Health Lessons from the Bhopal Chemical Disaster." *Journal of the American Medical Association* (December 5, 1990): 2795.

"Union Carbide's Bhopal Bill." *Business Week* (February 27, 1989): 40.

VIII

Medical Disasters

Diethylstilbestrol (DES)

United States
1940–79

Background

In 1971 the *New England Journal of Medicine* reported on seven cases of a rare form of vaginal cancer—adenocarcinoma—in girls and young women. This report followed others, all startling because of the high numbers of the cases and because they were adolescent girls and women no older than their early twenties. Prior to this "epidemic," the few cases of adenocarcinoma reported in the medical literature had all occurred in postmenopausal women over the age of 50. For the first time, however, a common link had been found. The *New England Journal* reported that for these seven young women, the mothers had all taken diethylstilbestrol (DES) during their pregnancies.

Estrogen was the name given to the biologically active substance first extracted from women's ovaries in 1912. The first nearly pure estrogen was prepared in the late 1920s by Edward Doisy, an American who later won the Nobel Prize in medicine.

A synthetic estrogen prescribed for decades to prevent miscarriage, DES produces cancers and reproductive abnormalities in children of women who took it.

DES is first synthetic estrogen

DES was the first of the synthetic estrogens. It was synthesized in 1938 by Sir E. Charles Dodds, a British biochemist. The chemical compound behaves physiologically like naturally occurring estrogens. However, its chemical structure, which is quite different from the naturally occurring hormone, is closely related to some of the cancer-causing substances found in coal tar. But DES was easy to make, inexpensive, and convenient—it could be administered orally. Natural estrogens could not.

Because estrogen deficiencies were recognized as causing abnormal conditions, the public and medical community were intensely interested in the synthetic hormone. George Smith and Olive Smith, a husband-and-wife team of respected Harvard researchers, were investigating why pregnancies in some women ended in fetal death or premature delivery. In 1934 the Smiths discovered that a sharp decline in estrogen levels occurred in the urine of pregnant women at least four weeks before the onset of their pregnancy complications.

DES was officially approved by the U.S. Food and Drug Administration (FDA) in 1940, one of the first major drugs to be approved under the revised Food, Drug, and Cosmetics Act of 1938. The revised legislation resulted from a tragedy that occurred in September and October 1937, when the drug Elixir Sulfanilamide killed at least 73 people. The elixir contained sulfanilamide dissolved in the poison diethylene glycol, which caused severe liver and kidney damage. The 1938 legislation required drugs to be proved safe before they could be marketed.

Proving long-term safety of drugs is not yet required by law

While drugs were required to be safe, the concepts of long-term safety and effectiveness had not yet evolved. Long-term animal toxicity studies were not yet required. And, until the Kefauver-Harris Act was added in 1962 to the Food, Drug, and Cosmetics Act, there was no requirement at all, not even a weak one, for drugs to demonstrate proven effectiveness.

The surge in DES use began in the late 1940s, when the Smiths published papers in the *American Journal of Obstetrics and Gynecology*. Olive Smith's 1948 paper, considered a landmark, claimed that DES could increase the chance for a successful pregnancy in women who had miscarried before or who had high blood pressure or diabetes, and that it might also prevent complications that occur late in pregnancy. This latter claim launched the wholesale use of DES as a prophylactic measure—that is, as a drug that could prevent complications of pregnancy. In many cases the drug was prescribed for women who had no symptoms or other indication that they required treatment.

At the same time, DES was becoming a popular drug among animal breeders. The drug made animals grow to marketable weight faster and on less feed. In 1947 the U.S. Department of Agriculture (USDA) approved implantable DES pellets to be inserted under the neck skin of chickens.

Details of the Disaster

The DES disaster is unusual because of the long interval—more than a generation—between the cause and its effects. The fact that a drug could have such long-delayed side effects in large part created the tragedy. Perhaps not surprisingly, laws and regulations devised with more conventional hazards in mind failed to function effectively either to prevent or contain the disaster.

Dieckmann study shows DES to be ineffective

DES use began to diminish in 1953 with the publication of two independent investigations conducted to study DES effectiveness. Both were controlled, double-blind trials in which neither patient nor physician knew whether the prescribed pill was really the medication or a placebo—a harmless pill containing no medication. Both studies showed that DES did not reduce miscarriages, premature births, or other late-pregnancy complications. Instead they showed that the synthetic hormone slightly increased the number of miscarriages. The larger of the two studies, begun in 1950 by William Dieckmann of the University of Chicago, involved some 2,000 women and produced evidence that DES-treated women had smaller babies and twice as many miscarriages.

The Dieckmann study was a turning point but not the death knell for DES. Sales of the drug peaked in 1953 and started gradually declining, primarily because of unfavorable reports that appeared regularly in medical journals. The drug was still legally marketed and widely prescribed. There was no law against marketing ineffective drugs, and drug companies were neither required to publicize nor even disclose that a drug was ineffective. No further developments occurred to change pregnancy-related use of DES until 1971.

DES approved for use in animals

Meanwhile, DES was being used in animals as a feed additive and as implantable pellets. Then, in 1950—three years after the USDA approved the use of implantable DES pellets in chickens—mink ranchers sued the USDA. The ranchers claimed that their minks became sterile after eating DES residues left in the heads and necks of treated chickens. Despite this suit, the USDA in 1954 extended the use of DES to cattle, sheep, and hogs, with the stipulation that treatment be stopped 48 hours before slaughter to prevent drug residues in the edible meat.

The claim that DES might prevent complications occurring late in pregnancy launched its wholesale use. In many cases, the drug was prescribed for women who had no symptoms or other indication that they required treatment.

DES enters human food chain through animal flesh

Nevertheless, traces of DES continued to be found in edible animal tissues. With DES known to cause cancer in animals, there was an outcry of concerns about food safety. This spurred the 1958 passage of the Delaney Amendment to the Food, Drug, and Cosmetic Act. This amendment flatly prohibited the use of any additives in human food that had been shown to cause cancer in either animals or humans. The Delaney amendment applied to additives put directly in foods as well as indirectly, as a result of the diet of an edible animal. However, the FDA argued that the Delaney amendment could not be applied retroactively to products whose use had already been approved (the so-called grandfathered products).

Improved analytic techniques detected dangerous levels of DES in chickens in 1959, and the use of DES in poultry was banned. However, because residues were not detected in beef, use of the drug in cattle continued—and grew steadily. By 1970, 75 percent of the beef cattle in the United States were fed DES.

DES linked to cancer in 1971—not banned until 1979

In spring 1971 Arthur Herbst and his colleagues at Massachusetts General Hospital published their paper linking DES with vaginal cancer. Seven months later—in November 1971—the FDA ruled that drug companies were required to warn against the use of DES during pregnancy. Not long after, it came to light that university health clinics were dispensing DES as a morning-after contraceptive, to induce miscarriage in women after unprotected sexual intercourse. DES had never been approved for this purpose.

Another 1971 revelation was that DES residues had been found in beef as early as 1966. Meats tested by established methods and declared safe were found to have traces of DES when they were subjected to a newly introduced method of detection. Since no amount of a cancer-causing agent, however small, is considered permissible, the inescapable conclusion was that there was no way to avoid DES residues in meat short of banning DES entirely. Yet the controversy and battles continued through 1979, when DES use in food-producing animals was finally banned.

Non-DES hormones—banned in Europe—can be used in U.S. animal feed

Today DES is banned worldwide as a growth promoter in cattle and poultry, but other naturally occurring hormones have taken its place. By some estimates U.S. farmers save between $2 billion and $4 billion in feed costs every year by using hormones. Although the prevailing expert opinion in the United States is that the use of such products does not produce hormone levels in edible tissues that are distinguishable from normal levels in the animals, cattle hormones of any kind are illegal in Europe.

Produces cancer and anatomical abnormalities in children of users

DES has been studied extensively in humans since 1971, and the results of these studies are pessimistic. It became clear in the late 1970s that DES daughters were subject to a higher incidence of vaginal cancer as well as a host of anatomical and functional irregularities of the reproductive tract: vaginal adenosis, T-shaped uteri, cockscomb cervixes, hooded cervixes, cervical ridges, and vaginal or uterine cysts. When these young women reached childbearing age, they encountered high rates of infertility, ectopic (tubal) pregnancies, and premature births.

Sons of DES mothers also show effects: they have a higher-than-average rate of infertility, testicular cancer, and other noncancerous anatomical abnormalities. The DES mothers themselves, decades after taking the drug, may be especially prone to vaginal adenocarcinoma, breast cancer, and cancer of the uterus.

Impact

Today the federally approved labeling for DES carries a boxed warning that use of estrogens increases the risk of cancer of the endometrium (the lining of the uterus) and that estrogens should not be used during pregnancy. The label also carries an unboxed warning that DES should not be used as a morning-after contraceptive. Despite such cautions and the consequent risk of birth defects if the method fails, DES is widely used for morning-after contraception. The principal approved uses of DES in women are for relief of symptoms of menopause and for treatment of a few specific conditions.

Grandchildren of DES users are third generation affected

Diethylstilbestrol was first marketed in 1941 by Eli Lilly & Company, a pharmaceutical firm that manufactured and sold about 75 percent of the DES consumed in the United States. The remaining 25 percent of the market went to some 20 other drug companies, including American Pharmaceutical Company, Rexall Drug & Chemical Company, E. R. Squibb & Sons, and Upjohn Company. Between 1947 and 1971 an estimated 6 million pregnant women took DES.

Numerous lawsuits have been brought against the drug companies. As of 1992 more than 1,000 DES cases had been filed nationwide, with Lilly named in 600 of them. About 10 percent of the cases involve so-called third-generation claims—that is, claims on behalf of the grandchildren of women who took DES. Among other impacts, the DES experience has changed the practice of American law, allowing plaintiffs to sue all manufacturers of DES when they are unable to discover which brand their mothers took.

In October 1991 a New York State jury handed down a $12.2 million verdict against Eli Lilly in a case involving a woman whose cancer and infertility were linked to her mother's DES use. It was the largest award in a DES case and also marked the first time a DES spouse was awarded damages. The plaintiff's husband won $550,000 in addition to the settlement awarded to his wife. The jury was in the process of deciding whether the company should also pay punitive damages when the two sides settled the case out of court, and terms of the final settlement were kept confidential. The case is expected to trigger drug companies to settle hundreds of other suits over the antimiscarriage drug.

From 40 to 60 percent of DES daughters have some form of benign abnormality in their reproductive organs, according to the National Cancer Institute. This puts them at higher risk for premature or protracted labor, which can result in problems at birth. There is an "indirect" link between the drug and third-generation babies being born with serious handicaps caused by uterine and cervical abnormalities in the DES daughters. Researchers currently claim that evidence does not exist that DES-induced defects might be passed down genetically to the third generation. However, a growing number of DES daughters, advocates, and medical experts are calling for more research into possible direct and indirect third-generation effects.

Where to Learn More

Apfel, R. J., and S. M. Fisher. *To Do No Harm: DES and the Dilemmas of Modern Medicine.* New Haven, CT: Yale University Press, 1986.

Dieckmann, W. J., M. E. Davis, L. M. Rynkiewicz, and R. E. Pottinger. "Does the Administration of Diethylstilbestrol during Pregnancy Have Therapeutic Value?" *American Journal of Obstetrics and Gynecology* (November 1953): 1062.

Epstein, S. S. *The Politics of Cancer.*New York: Anchor Press, 1979.

Fenichell, S., and L. Charfoos. *Daughters at Risk: A Personal DES History.* New York: Doubleday, 1981.

Herbst, A. L., H. Ulfelder, and D. C. Poskanzer. "Adenocarcinoma of the Vagina: Association of Maternal Stilbestrol Therapy with Tumor Appearance in Young Women." *New England Journal of Medicine* (April 22, 1971): 878–881.

McGrory, M. "The Lingering Pain of DES." *Washington Post* (April 23, 1992).

Meyers, R. *D.E.S.: The Bitter Pill.* New York: Seaview/Putnam, 1983.

Orenberg, C. L. *DES: The Complete Story.* New York: St. Martin's Press, 1981.

Schell, O. *Modern Meat: Antibiotics, Hormones, and the Pharmaceutical Farm.* New York: Random House, 1984.

Seaman, B., and G. Seaman. *Women and the Crisis in Sex Hormones.* New York: Rawson Associates, 1977.

Smith, O. W. "Diethylstilbestrol in the Prevention and Treatment of Complications of Pregnancy." *American Journal of Obstetrics and Gynecology* (November 1948): 821.

Thalidomide

More than 40 countries
1950s–1960s

Widespread use
of an improperly
tested drug
causes ghastly
birth defects in
thousands of
babies, half of
whom die within
weeks of birth.

Background

In the early 1960s, Thalidomide, an over-the-counter (nonprescription) pharmaceutical, was a popular sedative and antinausea drug marketed in more than 40 countries. It was found to be responsible for causing terrible birth defects in thousands of babies whose mothers had taken the drug during pregnancy to prevent morning sickness.

Some of the babies were born without arms or without legs. Others were born blind and deaf or with heart defects or intestinal abnormalities. Some of the babies were nothing more than a trunk with an eyeless, earless head mounted on it. Some were mentally retarded; most were of normal intelligence. It is thought that Thalidomide affected more than 10,000 babies around the world. This tragedy led directly to stronger laws regulating the development and sales of prescription and nonprescription drugs on four continents and to vastly improved testing of new drugs in the United States.

How new drugs used to be developed

Historically, new pharmaceuticals have been developed by trial and error, the so-called "suck-it-and-see method." Organic chemists synthesize new chemical compounds, then pharmacologists (scientists who study the effects of drugs) evaluate them for beneficial effects on lab animals. If a compound is found to have a desirable effect—a beneficial biochemical change—in the laboratory animals, the study is then expanded to human volunteers and finally to the general population.

Some of the babies born to mothers who had taken Thalidomide were born without arms or without legs, others with heart or intestinal defects. It is estimated that more than 10,000 babies worldwide were affected by the drug.

Even when performed correctly by dedicated scientists under favorable conditions, this method does not always result in the development of safe and effective drugs. When the process is compromised, either by incompetent scientists or by unscrupulous executives who want to rush a new drug to market, disaster can result.

Thalidomide was invented in West Germany

Thalidomide was originally developed in 1954 by Chemie Grünenthal, a West German pharmaceuticals company. Until then Grünenthal derived most of its revenues by marketing, under license, pharmaceuticals developed in other countries. But the profit margin on marketing the pharmaceuticals of other companies was low, so Grünenthal's scientists (many of whom had no formal training in pharmacology) were under intense pressure to develop new drugs. These drugs could then be marketed in West Germany. Grünenthal would not have its income diminished by royalty payments, and it could turn around and license other companies to sell its drug in other countries.

In the case of Thalidomide, Grünenthal submitted the drug to limited animal studies and cursory clinical trials on human beings. Then the West German pharmaceutical company rushed the drug to the market.

Thalidomide is wildly successful

By May 1961 Thalidomide was the most successful product in the history of the Grünenthal corporation. It accounted for more than half of Grünenthal's gross revenues and was by far the most popular over-the-counter sedative in West Germany. Grünenthal marketed Thalidomide as a cure for everything from nervous exhaustion and the flu to morning sickness. They took the position that Thalidomide was safe, completely nontoxic, and without side effects.

Other companies want a piece of the (profit) pie

Pharmaceutical companies throughout the world watched Grünenthal's success with Thalidomide. They lined up to get authorization to manufacture the lucrative drug. Licenses were granted to pharmaceutical companies in Britain, Japan, France, Italy, Australia, and the United States, the world's largest market for over-the-counter remedies.

Details of the Disaster

There occurred a sudden surge in birth defects in every nation of the world where Thalidomide was marketed. For more than six years Thalidomide was sold without anyone questioning its connection to the abnormalities that began to be seen. Before Thalidomide was suspected to be the cause, nearly 4,000 babies, mostly in West Germany, were born with ghastly birth defects. First to make the connection was an Australian obstetrician, Dr. William McBride, who delivered three babies in the first six months of 1961 with nearly identical birth defects. The three babies were born with no radius bone (the shorter of the two bones in the forearm) in either arm and with a bowel atresia (their lower intestine had no opening to the anus). All three babies died shortly after birth.

Australian doctor tries to get Thalidomide off the market

Dr. McBride researched the medical journals for information on bowel atresias and on birth defects in general. What he read led him to believe that the abnormalities he had seen in his practice were caused by an adverse drug reaction. He examined the hospital records of the three women who had given birth to the deformed babies—and found that all three women used Thalidomide during their pregnancies—and that it was the only drug any of them had taken.

This seemed to him to be compelling evidence that Thalidomide caused the birth defects and that it should be withdrawn from use until further studies could be conducted. Dr. McBride immediately notified the local distributor of Thalidomide that the drug caused birth defects, but he got no response. He also submitted a paper to the *Lancet,* one of the world's most prestigious medical journals, detailing his theory about Thalidomide. The *Lancet* refused to publish his paper. It would be another six months before Dr. McBride would be vindicated. Thousands more babies would die and thousands more babies would be born without arms and legs before Thalidomide would be withdrawn from the market.

American company hands out 2 million pills, then seeks FDA approval

Shortly before Dr. McBride delivered his first Thalidomide baby, Richardson-Merrel, the American licensee of Thalidomide and the maker of Vicks VapoRub, distributed samples of the drug even before submitting its application to the U.S. Food and Drug Administration (FDA). Richardson-Merrel had big plans for this over-the-counter marketing wonder. It saw to it that more than 2 million Thalidomide pills got into the hands of more than 1,200 doctors. They in turn dispensed the drug to

Defects to the eyes and ears occur if the mother takes Thalidomide between the 20th and 25th day of gestation, arms are affected between the 26th and 30th day, and legs between the 31st and 35th day.

approximately 20,000 patients. Then Richardson-Merrel submitted its application for FDA approval.

FDA refuses to approve the drug

Throughout 1961 Richardson-Merrel saw its application to market Thalidomide in the United States rejected on six different occasions by Dr. Frances Oldham Kelsey of the FDA. Almost single-handedly Dr. Kelsey faced down the multi-billion-dollar company, which was exerting strong pressure on her and her superiors to approve its Thalidomide application. It seemed the company would go to any length to get approval, even threatening Dr. Kelsey with a libel suit. Only after Thalidomide was shown to cause birth defects in West Germany and was withdrawn from the West German market did Richardson-Merrel withdraw its application for FDA approval.

Except for the courage of this single doctor, the Thalidomide disaster could have been many times worse. President John F. Kennedy awarded Dr. Kelsey the Distinguished Federal Civilian Service Medal for her courage. This medal is the highest award given to civilians in government service.

Biological Effects of Thalidomide

It seems logical that any drug to be marketed as a cure for pregnancy-induced morning sickness—as Thalidomide was—should be evaluated to see if it crosses the placental barrier. The placenta is the blood-rich organ through which the fetus draws nutrients and oxygen from the mother. The placenta protects the fetus from substances present in the mother's blood by limiting contact—there is no actual blood-to-blood contact between a mother and her fetus. Many harmful substances cannot cross the placenta from mother to fetus. Others can.

Placental barrier tests were not conducted

The original developer and marketers of Thalidomide—Chemie Grünenthal of West Germany, Distillers Company (Biochemicals) of the United Kingdom (the British distributor), and Richardson-Merrel of the United States—never produced such a study. It was not until after Thalidomide was withdrawn from the market as a suspected teratogenic (a substance that causes fetal malformations) that Grünenthal performed this vital experiment. Using a radioactive-tagged sample of Thalidomide,

Grünenthal scientists tested the drug in pregnant mice and found that it crossed the placental barrier.

Later on it would be learned that no laboratories could ever reproduce even one of Grünenthal's animal studies. And as for its clinical trials with human subjects, none used the standard double-blind testing—in which neither the doctor nor the patient knows if the patient is receiving the drug or a harmless placebo.

Birth defects occur in early pregnancy

It is still not understood how Thalidomide damages the developing fetus after it crosses the placental barrier. One theory is that the drug prevents the fetal blood vessels from forming properly. Blood clots form rather than blood vessels, resulting in tissue death in the areas affected. Depending on when during the fetal development cycle the drug is ingested, the effects can be wildly different. Defects to the eyes and ears occur if the mother takes Thalidomide between the 20th and 25th day of gestation, arms are affected between the 26th and 30th day, and legs between the 31st and 35th day. Morning sickness, for which Thalidomide was often prescribed, most often occurs during the first months of pregnancy. Thalidomide apparently had no effect on the fetus if the mother took the drug after the third month of pregnancy.

Recent beneficial side effect is discovered

Although the drug used in pregnant women during the early months of pregnancy had disastrous side effects, Thalidomide has another, actually beneficial side effect: it inhibits tumor necrosis factor (TNF). This side effect may prove useful in the treatment of AIDS. TNF is a natural substance that normally defends the body against cancer and infections, but for some reason it promotes the production of HIV viruses. Thalidomide seems to prevent HIV formation. Some AIDS patients who have taken Thalidomide have gained weight and no longer experience fevers. Although researchers do not consider Thalidomide a potential cure for AIDS, they believe the drug may slow the disease's progress by inhibiting the production of HIV viruses in infected cells.

Impact

There were fewer than a dozen known cases of Thalidomide-deformed births in the United States. The credit for so few incidents was attributed directly to the FDA, the federal agency formed to regulate new pharmaceuticals.

The United States created the FDA in 1927 in response to a major over-the-counter drug disaster. S. E. Massengill Company of Tennessee produced and marketed an untested sore-throat medication that killed more than 100 people, including many children. Until the Thalidomide disaster, the FDA was charged with enforcing the Food, Drug, and Cosmetics Act of 1938. After Thalidomide, that legislation was modified by the the Kefauver-Harris Act of 1962.

FDA given expanded powers

The Kefauver-Harris Act greatly strengthened the FDA. It required pharmaceutical companies to prove that drugs were safe *and* effective (prior to Thalidomide, pharmaceutical companies only had to show a drug was safe). The law also widely expanded the FDA's powers to regulate the development and testing of new drugs.

Drug companies had to get more scientific

Prior to Thalidomide, many pharmaceutical companies used their clinical trials more as a marketing scheme than as a genuine scientific study of a drug's safety and effectiveness. In the case of Thalidomide, Richardson-Merrel had given out more than 2 million free samples of Thalidomide prior to ever submitting the drug to the FDA for approval. No records were kept on who took the drug or any side effects they might have experienced. No one recorded the numbers and types of patients who were taking the drug. There was no requirement for participating doctors to inform their patients that they were prescribing an experimental drug.

The pharmaceutical companies were not even required to report all of their data to the FDA. There were known incidents of pharmaceutical companies holding back damaging information—in 1963 three executives of Richardson-Merrel were fined $80,000 and placed on probation for falsifying and withholding test data on an anticholesterol drug (this drug was also withdrawn from the market).

Tests had to be reproducible

After Thalidomide, the pharmaceutical companies were required to show that the doctors testing their new drugs were qualified and that the tests being conducted were safe, controlled, and reproducible. They also had to prove that no person was given a drug without being informed that the drug was experimental. The pharmaceutical companies were required to report all of the results of all of their tests, not just the ones that showed their new drugs in a favorable light.

Since the Thalidomide disaster, all drug trials in the United States are required to study the effects of the drug on fetuses. Subsequent Thalidomide testing showed that the drug did not lead to deformities in rat fetuses, but deformities nearly identical to those found in human beings showed up in rabbit fetuses. For this reason, every new drug tested in the United States is now evaluated for its ability to cross the placental barrier and is then tested in multiple animal species prior to being given to human beings.

Conflict of interest rules adopted

Many avenues for abuse have been cut off, and as new issues arise, new rules are adopted to protect the FDA from undue influence from the pharmaceutical industry. For example, in the early 1960s the head of the FDA's antibiotics division accepted more than $250,000 from pharmaceutical companies for writing promotional articles. Conflict of interest rules were subsequently passed forbidding FDA employees from accepting money from the drug industry.

The United States has the safest pharmaceutical industry in the world due to people like Dr. Frances Kelsey, whose moral courage under extreme pressure still inspires FDA doctors and scientists.

Where to Learn More

"Abortion and the Law." *Time* (August 3, 1962): 30.

"The Case for Thalidomide." *Time* (April 27, 1992): 23.

"Drug Scare Spurs Review of Safeguards." *Business Week* (August 4, 1962): 70–71.

Ferrara, J. L. M., and H. J. Deeg. "Graft-versus-Host Disease." *New England Journal of Medicine* (April 1991): 667–674.

Gorman, C. "An Old New Drug for AIDS." *Time* (July 12, 1993): 49.

Knightley, P., H. Evans, E. Potter, and M. Wallace. *Suffer the Children: The Story of Thalidomide.* New York: Viking Press, 1979.

Lambert, E. C. *Modern Medical Mistakes.* Bloomington: Indiana University Press, 1978.

Lear, J. "The Unfinished Story of Thalidomide." *Saturday Review* (September 1, 1962): 35–43.

Melville, A., and C. Johnson. *Cured to Death: The Effects of Prescription Drugs.* New York: Stein & Day, 1982.

"Tough Drug Law Coming." *Business Week* (August 11, 1962): 37–38.

Will, J. "The Feminine Conscience of the FDA: Dr. Frances Oldham Kelsey." *Saturday Review* (September 1, 1962): 41–43.

Silicone-gel implants

1960s–

Background

Over 1 million American women have received silicone-gel breast implants during the past 30 years. The silicone-gel models are more popular than their saline, or salt-water–filled cousins, because silicone produces a more lifelike result. Implants are installed surgically for various cosmetic reasons: to restore the size of a woman's breasts after a mastectomy (breast removal), to augment the size of a woman's breasts, and to correct problems resulting from birth defects or medical traumas. Up until January 1992, 120,000 to 150,000 women a year underwent this surgery. Occasional case studies of health problems associated with the implants appeared in medical journals over a period of many years. These studies cited no deaths connected with the use of silicone-gel implants, but they did implicate a variety of health problems.

Silicone-gel breast implants leak or rupture, causing inflammation, fever, arthritis, scleroderma (a hardening of the skin), and other health problems.

Silicone implants develop problems

When silicone-gel implants were first sold in the early 1960s, there was no federal regulation governing them. They came under U.S. Food and Drug Administration (FDA) jurisdiction in 1976. In 1988 the FDA put in place new regulations that required manufacturers of the implants to supply more safety information about their products by July 1991. Meanwhile, concerns about the safety of silicone-gel implants were gaining more publicity.

The implants were found to have several problems:

• Deterioration over time requires replacement.

Available in a variety of sizes, implants comprise a sac of silastic filled with silicone gel. While the sac itself is regarded as harmless, it is the inner gel that, according to opponents, sabotages the body's chemical defense system and causes multiple adverse reactions.

- Leakage occurs from the flexible sacs.

- Sacs rupture (conservative sources say the rupture rate is less than 1 percent; government sources claim the rate to be 4 to 6 percent).

Many medical products made of silicone since 1940s

Silicone, or polydimethylsiloxane, its scientific name, can appear in liquid, gel, or solid states—depending on the length of the molecules in the polydimethylsiloxane compound. Silicone has been used since World War II in a wide variety of medical products, including needles, syringes, joint prostheses (such as artificial hip joints), heart bypass technology (as in synthetic tubes used in the heart), the treatment of diabetes, hemodialysis (blood purification treatments), and test tube linings.

Breast implants made with silicone gel come in various sizes and have two main components: a sac often made of silastic, an elastic silicone rubber; and the silicone gel itself, which may vary in viscosity according to how firm the patient wants the implants to be. The silastic envelope is assumed to be biologically inert; that is, it does not react with the human immune system. It is thought to cause no adverse reactions. Silicone gel, however, will react with the body's chemical defense system and cause adverse reactions.

Leaking gel can produce life-threatening symptoms

Various companies manufactured silicone-gel implants, including Bristol-Myers Squibb Company, Dow-Corning, Minnesota Mining and Manufacturing Company, and Mentor Corporation. As early as 1974 Dow-Corning scientists discovered that the silicone-gel implants could react adversely with the body's autoimmune system, but the public did not find out for another 15 years. The U.S. attorney general discovered in a 1992 investigation that prior to 1987, Dow-Corning employees falsified data about the oven temperatures used in the process of manufacturing the implants. Dow-Corning manufactured the majority of the silicone-gel implants used in the United States.

Details of the Health Problems

Health problems that have been associated with silicone-gel implants include arthritis, inflammations, swollen joints, rashes, autoimmune system problems—which arise when the body tries to reject foreign objects, and scleroderma—a disease that causes the skin to become tight and leathery. The two cases that follow illustrate some of the reactions to silicone-gel implants.

Case history 1: Woman's temperature soars to 104°F

In the first case, a 32-year-old woman elected to have breast augmentation surgery in 1977. The day after the silicone-gel implants were put in place in an uneventful operation, the woman came down with a fever of 104°F (40°C) accompanied by drenching sweats. Four days after her operation, she began to experience unexplained pain in her fingers and toes. Eight days after surgery, the joints in her hands, knees, and ankles began to swell. On the tenth postoperative day, the woman felt pain throughout her abdomen, and her doctors discovered that both of her kidneys had enlarged. Yet all this time the incisions on her breasts were healing more or less normally. Furthermore, thorough physical examinations as well as tests of her blood, urine, and other fluids showed no cause for the pain or swelling. Nonetheless, the woman's decline continued. She found it more and more difficult to breathe, her kidneys began to deteriorate, and she became delirious.

On the eleventh day after her operation, her doctors decided to remove her breast implants because they could find no other explanation for her problems. The woman's condition improved sharply within 24 hours.

When doctors inspected the silicone-gel implants after removal, the silastic sacs looked normal. During the surgery to remove the implants, the doctors took urine, blood, and sputum(spit) samples from the patient, and they collected fluid from both breast pockets to try to find the cause of the symptoms. All of these samples showed the presence of silicone compounds.

Six days after the removal of the woman's implants, most of her symptoms subsided and she was discharged from the hospital. Three weeks later, further samples of blood and breast fluid showed no silicone present. After three months the woman had healed completely. After testing the woman extensively, the doctors could find no explanation for her rapid deterioration in health. Because her symptoms began shortly after augmentation surgery and because the symptoms subsided sharply after the implants were removed, her doctors could only conclude that the implants had caused the incident.

Case history 2: Woman develops scleroderma

The second case involved a 46-year-old woman who underwent three implant surgeries. She had silicone-gel breast implants in 1976 for cosmetic reasons. Her second operation occurred in September 1987, after her doctors discovered that one of her first implants had either ruptured or leaked silicone gel. For a while after her operation the woman seemed to show no ill effects. Then in November 1987 her face, hands, legs, and feet began to swell intermittently. She went to her doctor several times for treatment, which temporarily improved her symptoms. She continued to experience tightness in the skin of her face, hands, legs, and feet, then began to develop firm, discolored patches of skin on the front side of her lower legs. In July 1988 a skin biopsy showed that she had scleroderma, which, if left untreated, could begin to affect her internal organs and eventually lead to death. Another examination in August 1988 showed that, aside from the swelling and discolored patches of skin, her health seemed normal.

The woman's scleroderma symptoms, however, worsened, and the discolored patches of skin spread up her legs to her thighs. Her legs became so stiff that she could no longer run or participate in her aerobics classes. Due to the negative reports about silicone-gel implants being published in the media at the time, the woman decided to have her silicone implants removed and replaced with saline, or salt-water, implants. Two weeks after having the silicone implants removed, the woman felt much better and her symptoms had improved significantly. One year later she was completely recovered and resumed her running and aerobics classes.

More than 90 percent of silicone recipients are happy with them

Yet despite the number of case studies associating silicone-gel implants with health problems, there are far more examples of these implants that have been used without any complications. In fact, more than 90 percent of the women who have had silicone-gel implants say they are satisfied with them. A number of physicians have published studies of the effects of silicone-gel implants on women's health, and they have concluded that the implants are safe. Although negative reactions to the implants clearly occur, it is equally clear that such reactions are infrequent.

Impact

Three significant changes have resulted from the controversy over the safety of the silicone-gel implants:

- Most of the manufacturers of the implants have gotten out of the business.

Most of the manufacturers of the implants have gotten out of the business; several thousand lawsuits fill judicial dockets and will be tried or settled over many years to come; the FDA has undertaken a long-range, large-scale study to see how safe silicone-gel implants really are.

- Several thousand lawsuits fill judicial dockets and will be tried or settled over many years to come.

- The FDA has undertaken a long-range, large-scale study to see how safe silicone-gel implants really are.

In 1988 the Food and Drug Administration told manufacturers of silicone-gel implants that they had to present detailed safety information about their products to the FDA by July 9, 1991. The safety information presented to the FDA showed no clear link between silicone-gel implants and autoimmune disorders or other systemic diseases. Nevertheless, the lack of long-term safety data about the implants caused Dr. David Kessler, the commissioner of the FDA, to declare a 45-day moratorium on the sale of the silicone-gel implants beginning January 6, 1992.

Lawsuits win large awards

Kessler's moratorium received extensive coverage in the media, which exacerbated the legal and financial problems of the silicone-gel manufacturers that had begun some time before. In 1984 a San Francisco, California, jury found against Dow-Corning and awarded a Nevada woman $1.5 million in damages. The case was settled out of court for an undisclosed sum after Dow-Corning appealed the decision.

In December 1991 another San Francisco jury issued a $7.3 million verdict against Dow-Corning. At least part of the reason for the large judgment was the evidence showing that Dow-Corning had not warned doctors that Dow had linked ruptured silicone-gel implants with autoimmune diseases.

After Dr. Kessler's moratorium in January 1992, the volume of lawsuits increased substantially. By February 1993 an estimated 3,000 lawsuits related to silicone-gel implants had been filed in state and federal courts. Late in January 1993 a Texas state court awarded $25 million to a claimant against the Bristol-Myers Squibb Company. Bristol-Myers Squibb had closed its silicone-gel implant business in September 1991. Dow-Corning ceased silicone-gel breast implant manufacture in March 1992.

Implant manufacturers quit the business

In September 1993 Dow-Corning announced plans to create a $4.75 billion liability fund to compensate those injured by silicone-gel implants. Each victim could receive between $200,000 and $2 million; removal of the implants would also be covered. The fund would be paid into by those being sued—insurance companies, doctors, and manufacturers. The plan awaits approval by a federal judge. A settlement was also reached to end the manufacture of silicone-gel implants by U.S. companies by 1995. By mid-1993 only one company—the Mentor Corporation—was still in the business.

Despite the moratorium on the sale of the implants, the FDA will allow silicone-gel implant surgery for women who need to replace a device already in place and for women who need implants after a mastectomy.

Finally, one of the most significant outcomes of the implant controversy is that the FDA will conduct a long-term, large-scale, scientifically controlled study on a limited number of women to decide once and for all about the safety and effectiveness of silicone-gel implants.

Where to Learn More

Angell, Marcia. "Breast Implants: Protection or Paternalism?" *New England Journal of Medicine* (June 18, 1992): 1695–1696.

Burton, Thomas M. "Breast Implants Raise More Safety Issues." *Wall Street Journal* (February 4, 1993): B1, B8.

Fisher, Jack C. "The Silicone Controversy: When Will Science Prevail?" *New England Journal of Medicine* (June 18, 1992): 1696–1698.

Kessler, David A. "The Basis of the FDA's Decision on Breast Implants." *New England Journal of Medicine* (June 18, 1992): 1713–1715.

Sahn, Eleanor E., and others. "Scleroderma following Augmentation Mammoplasty." *Archives of Dermatology* (September 1990): 1198–1202.

Smart, Tim. "Breast Implants: What Did the Industry Know, and When?" *Business Week* (June 10, 1991): 94–98.

Uretsky, Barry F., and others. "Augmentation Mammoplasty Associated with a Severe Systemic Illness." *Annals of Plastic Surgery* (November 1979): 445–447.

Dalkon Shield intrauterine device

1970s

An intrauterine birth control device causes fatalities and causes spontaneous abortions, internal injuries, infertility, and birth defects.

Background

The Dalkon Shield was an intrauterine birth control device (IUD) sold from 1971 to 1975. A doctor would implant the device in a woman's uterus to prevent pregnancy. Right from the beginning problems began to occur. At least 20 deaths and thousands of cases of internal injuries have been blamed on the shield. Evidence indicates that its manufacturer, the A. H. Robins Company, knew about potential problems with the shield all along—but marketed it anyway and promoted it heavily. After widespread bad publicity, the company removed the product from the U.S. market in late June 1974. Robins suspended the sales of the Dalkon Shield outside of the United States in April 1975. In October 1984, after almost ten years of litigation, Robins finally began to urge doctors to remove shields still in use.

A flat piece of flexible plastic shaped like a crab, the Dalkon Shield measures about three quarters of an inch across. The five prongs on each side of the shield keep it in place in the uterus and keep the woman's body from spontaneously expelling it. The prongs also make the shield difficult to insert and remove, and they can puncture the lining of the uterus.

Tailstring allows woman to check it herself

The shield has a plastic string called a tailstring tied to its bottom portion. This helps a doctor to remove it and allows a woman to check that the shield is properly in place. The upper end of the tailstring is threaded through a small opening in the bottom part of the shield and knotted, and

Opponents of the Dalkon Shield (Karen Hicks, left, president of the Dalkon Shield Information Network) stand vigil outside a federal courthouse. They are waiting for the court to determine compensation to women injured by the device.

it stays in the woman's uterus. The string then extends through the cervix, or lower neck-like end of the uterus, and ends in the vagina.

Tailstring is a highway for germs

It turned out that tailstring construction facilitated infections in the uterus. The string was made of tiny nylon strands encased in a sheath. The ends of the sheath were not sealed shut, which allowed moisture and bacteria to travel up the sheath from the vagina into the uterus in a process called "wicking." In wicking, which can even occur vertically, a fluid is attracted through fibrous materials because of the surface tension of the fluid. In the case of wicking in the tailstring of the Dalkon Shield, bacteria traveled from the vagina to the uterus and caused pelvic inflammatory diseases (PID). PID can be fatal or life-threatening.

Hugh Davis, a doctor, and Irwin Lerner, an engineer, invented the Dalkon Shield. Davis tested the shield for 12 months on 640 women. He published the results of his test in the February 1970 issue of the *American Journal of Obstetrics and Gynecology.* Davis claimed an annual pregnancy rate of 1.1 percent, which was substantially better than any other intrauterine device.

Research was flawed

But Davis's study had important flaws:

1. Too few experimental subjects were included.
2. Only eight patients wore the device in the first month of the study.
3. Contraceptive foam was used during the first few months after shield insertion.
4. Too short a time (only three days) lapsed after the 12-month test period before data was finalized (any subsequent pregnancies that occurred in the 640 test subjects were not reported in the study).
5. No disclosure of Davis's financial stake in the success of the shield.

Robins buys shield from inventors

A. H. Robins Company, a Fortune 500 pharmaceuticals corporation based in Richmond, Virginia, purchased the rights to the Dalkon Shield from its inventors, Davis and Lerner, in June 1970.

In the days before it purchased the shield, Robins learned that Davis had additional data that modified his earlier claims. The additional data reported the pregnancy rate from the shield as a minimum of 5.3 percent. Robins also found out that no data existed regarding long-term safety of the device.

A few days after Robins bought the rights to the shield, Lerner informed Robins that the tailstring might be a problem. About two weeks later a Robins executive sent a memo to 39 Robins executives, advising that the shield had a "wicking" tendency that should be looked into carefully.

Robins withheld information about shield's problems

Despite the safety problems, Robins began to sell and promote the Dalkon Shield vigorously in January 1971. Robins gave its sales staff only the 1.1 percent pregnancy rate, and it advertised the shield in medical journals as being "trouble free" and "safe and effective." Eventually an estimated 3.6 million women used the shield.

Details of the Health Problems

Women reported that the process of having the Dalkon Shield inserted was painful; there were reports of fainting, agonizing cries, and vaginal bleeding. Later complications included pelvic infections, infertility, perforations of the uterus, scarred internal reproductive organs, spontaneous abortions caused by infections, and premature or stillborn babies. Over 5 percent of the women who used the shield to avoid pregnancy became pregnant anyway.

Two case histories

Pam Van Duyn had the shield inserted in October 1977. A few hours later she developed a high fever. The first doctor she visited thought she had food poisoning and gave her penicillin and other antibiotics. After five days in agony, Van Duyn saw a gynecologist who diagnosed her as having acute pelvic inflammatory disease (PID). A runaway bacterial infection, PID can affect a woman's ovaries, uterus, and other pelvic organs. Left unchecked, PID can be fatal. The infection can also scar the fallopian tubes, which carry the egg down to the uterus. The scarring can make a woman infertile.

In June 1970 A. H. Robins Company purchased the rights to the Dalkon Shield from its inventors. Despite a lack of any data on the long-term safety of the device, Robins began to sell and promote the Dalkon Shield vigorously in January 1971. An estimated 3.6 million women used it.

While examining Van Duyn, the doctor found a large mass the size of an orange attached to her left ovary. A lab culture identified the bacteria and indicated which antibiotics would be effective. Hospitalized, Van Duyn responded to the antibiotic, which controlled the infection. Nine days later Van Duyn left the hospital, but she had to rest in bed for several weeks and felt weak for months after that. Van Duyn's pelvic mass did not abate, and she had a low-grade, chronic infection that flared up from time to time. In 1983 Van Duyn had her left ovary and fallopian tube removed, and in 1984 she had surgery to remove the scarring in her right fallopian tube. Under tremendous stress from the ordeal, she and her husband separated, though they reunited later.

Peggy Mample had a Dalkon Shield inserted in autumn 1971 and reported an immediate worsening of her menstrual periods, which became very heavy and painful. In March 1972 she became pregnant despite the presence of the shield. To minimize the chances of a miscarriage, she decided to leave the shield in place. Seven months later Mample had an inflammatory reaction to the shield, which induced labor. Her daughter, Melissa, was born two months prematurely. The baby girl had dislocated hips and hyaline membrane disease, a respiratory condition that can lead to retardation because of oxygen starvation. Melissa gained strength and was released from the hospital, but a year later Mample learned that Melissa had a form of cerebral palsy that paralyzed the muscles of her legs. Melissa could walk only with the aid of leg braces and a walker. Though Melissa's mental abilities seemed unaffected, her motor skills were so poor that she was confined to a wheelchair or had to drag herself along the floor. The stress was more than Mample and her husband could bear, and they were divorced. Mample remarried and later gave birth to two healthy sons.

Impact

Some women filed suits against A. H. Robins over the Dalkon Shields. Many of the early cases were settled out of court for relatively small sums—such as $10,000. But in February 1975 a Kansas woman won $85,000 in damages against Robins. That released a flood of complaints. By the middle of 1975, more than 500 U.S. legal cases demanding $500 million in damages were pending against A. H. Robins. Some women won over $1 million each in their cases against the company, and one Kansas woman in 1985 won $8.9 million in damages. Eventually it was discovered that lawyers for Robins had destroyed valuable evidence. This happened at the request of the company's chief legal counsel and with the knowledge of the company's president. Some of the evidence had been secretly photocopied by a Robins lawyer. It showed definitively that Robins knew all along about the Dalkon Shield's potential to damage women.

Robins files for bankruptcy and schemes to keep control of company

By the mid-1980s, management at A. H. Robins expected that more large financial judgments would be assessed against the company. In a move to cut its losses, the company—controlled by E. Claiborne Robins Sr. and E. Claiborne Robins Jr.—filed for bankruptcy on August 21, 1985. The Robins family hoped to cope with the losses and ultimately maintain company control, but during the long bankruptcy process, Robins became a target for takeover by other companies.

Robinses get the boot

On December 15, 1989, Robins officially merged with American Home Products. For their family-owned share of the Robins stock, E. Claiborne Robins Sr. and Jr. received $385 million, a tax-free exchange of stock. Shortly after the takeover, both Robinses were told that American Home Products no longer required their services.

The bankruptcy process lasted over four years, during which over 325,000 claims against Robins were filed. Of these the federal courts allowed 195,000. Ultimately $2.7 billion was paid into a trust to settle the claims of women who used the Dalkon Shield. It will probably be well into the twenty-first century before the litigation is over.

Where to Learn More

Alexander, Charles P. "Robins Runs for Shelter." *Time* (September 2, 1985): 23.

Chakravarty, Subrata N. "Tunnel Vision." *Forbes* (May 21, 1984): 214, 218.

Culliton, Barbara J., and Debra S. Knopman. "Dalkon Shield Affair: A Bad Lesson in Science and Decision-Making." *Science* (September 6, 1974): 839–841.

Mintz, Morton. *At Any Cost: Corporate Greed, Women, and the Dalkon Shield.* New York: Pantheon Books, 1985.

Perry, Susan, and Jim Dawson. *Nightmare: Women and the Dalkon Shield.* New York: Macmillan, 1985.

Sobol, Richard B. *Bending the Law: The Story of the Dalkon Shield Bankruptcy.* Chicago, IL: University of Chicago Press, 1991.

Bibliography

Barclay, Stephen. *The Search for Air Safety: An International Documentary Report on the Investigation of Commercial Aviation Accidents.* New York: William Morrow, 1970.

Bignell, Victor, Geoff Peters, and Christopher Pym. *Catastrophic Failures.* Bristol, PA: Open University Press, 1977.

Bishop, R. E. D. *Vibration.* Winchester, MA: University Press, 1965.

Blockley, D. E. *The Nature of Structural Design and Safety.* Ellis Horwood, 1980.

Canning, John, ed. *Great Disasters.* Stamford, CT: Longmeadow Press, 1976.

Collins, J. A. *Failure of Materials in Mechanical Design: Analysis, Prediction, Prevention.* New York: Wiley, 1981.

Cornell, James. *The Great International Disaster Book.* New York: Scribner's, 1976.

Davis, Lee. *Man-Made Catastrophes: From the Burning of Rome to the Lockerbie Crash.* New York: Facts on File, 1993.

Ebert, Charles H. V. *Disasters: Violence of Nature, Threats by Man.* Dubuque, IA: Kendall/Hunt, 1988.

Eddy, Paul, Elaine Potter, and Bruce Page. *Destination Disaster: From the Tri-Motor to the DC-10; The Risk of Flying.* Quadrangle, 1976.

Editors of Encyclopaedia Britannica. *Catastrophe! When Man Loses Control.* New York: Bantam/Britannica Books, 1979.

Feld, Jacob. *Construction Failures.* New York: Wiley, 1968.

———.*Lessons from Failures of Concrete Structures.* Detroit, MI: American Concrete Institute, 1964.

Florman, Samuel C. *Blaming Technology: The Irrational Search for Scapegoats.* New York: St. Martin's Press, 1981.

Ford, Daniel. *O-Rings and Nuclear Plant Safety: A Technological Evaluation.* Washington, DC: Public Citizen, Critical Mass Energy Project, 1986.

Frank, Beryl. *Great Disasters of the World.* Austin, TX: Galahad Books, 1981.

Godrey, Edward. *Engineering Failures and Their Lessons.* Privately printed, 1984.

Godson, John. *Unsafe at Any Height.* New York: Simon & Schuster, 1979.

Gordon, J. E. *The New Science of Strong Materials; or, Why You Don't Fall through the Floor.* 2d ed. New York: Penguin Books, 1976.

————. *Structures; or, Why Things Don't Fall Down.* New York: Da Capo Press, 1981.

Great Britain Navy Department Advisory Committee on Structural Steels. *Brittle Fracture in Steel Structures.* Markham, Ontario: Butterworths Canada, 1970.

Guide to Investigations of Structural Failures. Washington, DC: Federal Highway Administration, 1980.

Hammond, Rolt. *Engineering Structural Failures: The Causes and Results of Failure in Modern Structures of Various Types.* Odhams Press, 1956.

Hertzbert, R. W. *Deformation and Fracture Mechanics of Engineering Materials.* New York: Wiley, 1976.

Janney, Jack R. *Guide to Investigation of Structural Failures.* New York: American Society of Civil Engineers, 1979.

Keylin, Arleen, and Gene Brown. *Disasters: From the Pages of the New York Times.* New York: Ayer, 1976.

Kletz, Trevor A. *What Went Wrong? Case Histories of Process Plant Disasters.* Houston, TX: Gulf, 1985.

Launay, A. J. *Historic Air Disasters.* London: Ian Allen, 1967.

LePatner, Barry B., and Sidney M. Johnson. *Structural and Foundation Failures: A Casebook for Architects, Engineers, and Lawyers.* New York: McGraw-Hill, 1982.

Lewis, Elmer Eugene. *Nuclear Power Reactor Safety.* New York: Wiley, 1977.

Mair, George. *Bridge Down: A True Story.* Briarcliff Manor, NY: Stein & Day, 1983.

McClement, Fred. *It Doesn't Matter Where You Sit.* New York: Holt, 1969.

————. *Jet Roulette: Flying Is a Game of Chance.* New York: Doubleday, 1978.

McKaig, Thomas K. *Building Failures: Case Studies in Construction and Design.* New York: McGraw-Hill, 1963.

Nash, Jay Robert. *Darkest Hours*. Chicago: Nelson-Hall, 1976.

Oberg, James E. *Uncovering Soviet Disasters*. New York: Random House, 1988.

Osgood, Carl C. *Fatigue Design*. 2d ed. Elmsford, NY: Pergamon Press, 1982.

Perrow, Charles. *Normal Accidents: Living with High-Risk Technologies*. New York: Basic Books, 1984.

Petroski, Henry. *Design Paradigms*. New York: Cambridge University Press, 1994.

——. *To Engineer Is Human*. New York: St. Martin's Press, 1985.

Ross, Steven S. *Construction Disasters: Design Failures, Causes, and Prevention*. New York: McGraw-Hill, 1984.

Salvadori, Mario. *Why Buildings Stand Up: The Strength of Architecture*. New York: McGraw-Hill, 1982.

Salvadori, Mario, and Matthys Levy. *Why Buildings Fall Down: How Structures Fail*. New York: Norton, 1992.

Serling, Robert J. *Loud and Clear: The Full Answer to Aviation's Vital Question, Are the Jets Really Safe?* New York: Doubleday, 1969.

Stewart, Oliver. *Danger in the Air*. New York: Philosophical Library, 1958.

Turner, Barry A. *Man-Made Disasters*. New York: Crane, Russak, 1978.

U.S. House of Representatives. Committee on Science and Technology. *Structural Failures: Hearings before the Subcommittee on Investigations and Oversight*. Washington, DC: Government Printing Office, 1984.

——. *Structural Failures in Public Facilities*. Washington, DC: Government Printing Office, 1984.

Webb, Richard E. *The Accident Hazards of Nuclear Power Plants*. Amherst: University of Massachusetts Press, 1976.

Whyte, R. R., ed. *Engineering Progress through Trouble*. Institution of Mechanical Engineers, 1975.

Index

Bold denotes entries.